Brain-Computer Interfaces for Human Augmentation

Brain-Computer Interfaces for Human Augmentation

Special Issue Editors

Riccardo Poli
Davide Valeriani
Caterina Cinel

MDPI • Basel • Beijing • Wuhan • Barcelona • Belgrade

MDPI

Special Issue Editors

Riccardo Poli
University of Essex
UK

Davide Valeriani
Harvard Medical School
USA

Caterina Cinel
University of Essex
UK

Editorial Office
MDPI
St. Alban-Anlage 66
4052 Basel, Switzerland

This is a reprint of articles from the Special Issue published online in the open access journal *Brain Sciences* (ISSN 2076-3425) from 2018 to 2019 (available at: https://www.mdpi.com/journal/brainsci/special_issues/Brain_Computer_Interface).

For citation purposes, cite each article independently as indicated on the article page online and as indicated below:

LastName, A.A.; LastName, B.B.; LastName, C.C. Article Title. *Journal Name* **Year**, *Article Number*, Page Range.

ISBN 978-3-03921-906-3 (Pbk)
ISBN 978-3-03921-907-0 (PDF)

Contents

About the Special Issue Editors

Riccardo Poli (Professor of Computer Science) is a full Professor in the School of Computer Science and Electronic Engineering of the University of Essex, where he is the coordinator of the Brain Computer Interfaces and Neural Engineering laboratory. Prof. Poli is a biomedical engineer (by first degree, Ph.D., and subsequent research) as well as an expert in genetic and evolutionary computation, and more generally machine learning and computational intelligence. He has written over 300 refereed publications and two books. According to Google Scholar, he has approximately 20,000 citations and an H-index of 61.

Davide Valeriani (Postdoctoral Research Fellow) is a computer engineer with research interests in the area of brain–computer interfaces (BCIs) for augmenting human performance. He received his Ph.D. from the University of Essex in 2017 with a thesis on collaborative BCIs for augmenting group decision making. Currently, he is a research fellow at Harvard Medical School, developing machine-learning models to make automatic diagnosis of speech disorders from multimodal neuroimaging data, as well as continuing his research in BCIs for decision making.

Caterina Cinel (Postdoctoral Research Fellow) is a researcher in cognitive neurosciences, with interests in the areas of multisensory integration, attention and memory, decision-making, brain–computer interfaces, and cognitive augmentation. She received her Ph.D. from the University of Birmingham (UK), and is currently an active member of the Brain-Computer Interfaces and Neural Engineering Lab at the University of Essex (UK).

Editorial

Brain–Computer Interfaces for Human Augmentation

Davide Valeriani [1], Caterina Cinel [2] and Riccardo Poli [2,*]

[1] Department of Otolaryngology, Massachusetts Eye and Ear, Harvard Medical School, Boston, MA 02114, USA; davide_valeriani@meei.harvard.edu

[2] Brain-Computer Interfaces and Neural Engineering Laboratory, School of Computer Science and Electronic Engineering, University of Essex, Wivenhoe Park, Colchester CO4 3SQ, UK; ccinel@essex.ac.uk

* Correspondence: rpoli@essex.ac.uk; Tel.: +44-1206-872338

Received: 21 January 2019; Accepted: 22 January 2019; Published: 24 January 2019

The field of brain–computer interfaces (BCIs) has grown rapidly in the last few decades, allowing the development of ever faster and more reliable assistive technologies for converting brain activity into control signals for external devices for people with severe disabilities. In recent years, however, the scope of BCIs has been extended from assistive technologies to neuro-tools for human cognitive augmentation for everyone. For instance, novel applications of BCIs have been proposed, enabling people to go beyond human limitations in sensory, cognitive, and motor tasks [1–4]. These include new and exciting paradigms, such as BCIs based on the brain activity of multiple people [5].

The aim of this special issue was to gather high-quality papers—including both reviews and reports on novel research—representative of the ongoing research in the area of BCIs for human cognitive augmentation. Twelve manuscripts were received through the open submission window, which went through a rigorous selection, peer review, and revision process, resulting in five papers being accepted for publication within the special issue. These papers are briefly described below.

One of the earliest BCI applications, the famous P300 matrix speller developed by Farwell and Donchin over three decades ago [6], provided a simple and practical way for restoring communication capabilities to the paralyzed. Since then, a large variety of spellers have been developed, which explore different paradigms, graphical user interfaces, neuroimaging techniques, and signals from the brain used to control the device. The paper by Rezeika et al. [7] in this special issue presents a thorough overview of the main EEG-based spellers that have been developed in the current decade (Jan 2010–Jan 2018). The authors propose a taxonomy based on the type of neural activity exploited: P300, steady-state visual-evoked potentials (SSVEP), motor imagery (MI), or hybrid. They further categorize the spellers based on operation, selection, stimuli modality, gaze dependency, and word prediction, also highlighting the need of keeping the final users in the loop when testing new BCIs. We hope this review will serve as a reference point for researchers interested in the area of BCI-mediated communication.

Given the importance of spellers in BCI research, it is not surprising that another paper in this special issue focuses on this. One of the most common limitations of BCI spellers is that they are typically tested with able-bodied users, but then fail when tested with locked-in patients. Tonin and colleagues [8] propose a novel BCI speller, that potentially enable patients in the complete locked-in state to express their thoughts, needs, and desires. This speller does not rely on letter-by-letter spelling. Instead, the speller is based on yes/no questions, aimed at gradually restricting possible interpretations and eventually allowing guessing the sentence that the patient would like to spell. The binary answers of the patient are decoded from his/her brain signals, recorded using functional neural infrared spectroscopy (fNIRS). Thanks to an artificial neural network and a binary decoding together with a sequence of questions, this BCI achieves higher accuracy than other BCI spellers.

Many BCI applications, starting from spellers, are based on event-related potentials (ERPs) recorded with EEG. It is, therefore, vital to be able to identify those ERPs (e.g., the P300) from the raw EEG signal recorded from the user's scalp. The third article of this special issue, by Ramele and colleagues [9], reviews the main methods used for detecting patterns in the EEG activity that could be used in a BCI. The authors compare different methods on both a pseudo-real dataset and the public dataset BCI competition II, both based on (again) a P300-based BCI speller. The authors conclude that fully-automated solutions for identifying such patterns are often suboptimal, and that hybrid systems, using both machine-learning algorithms and the experience of clinicians, may allow BCIs to reach higher accuracies.

The final two articles of this special issue focus on novel applications of BCIs for human augmentation. Nayak and colleagues [10] explore the possibility of detecting changes in human performance, as temperature changes in a work environment, from brain signals. In their study, they have monitored EEG, skin temperature, and heart rate while users were undertaking some office tasks of different difficulty level (i.e., arithmetic problem-solving and typing). They used the room temperature as an independent variable to change the performance of the users in the task, as people are more efficient when put in a comfortable environment. Then, they used neural and physiological signals separately to predict the performance of the user. Weak correlation was found between either the heart rate or the skin temperature and performance level. However, Nayak and colleagues found that EEG features in the power spectrum make good predictors of the performance level of the user. These findings could lead to the development of closed-loop, passive BCIs [11] able to monitor workers and adjust in real time the environmental conditions to maximize their performance.

The last article of this special issue proposes a novel paradigm for integrating humans and machines. In the future, it is very likely that many tasks will be performed by artificial intelligence (AI), but it is also extremely likely that in many other complex tasks there will be a tight integration between humans and AI devices. To achieve the latter, Marc Cavazza [12] proposes to use a BCI to keep the human in the loop, using his/her brain signals to influence the internal heuristic searches performed by the AI devices: the main computations are still performed by AI, with the human, however, being able to supervise the task. The BCI measures the variations of prefrontal asymmetry from a baseline and uses a mapping algorithm to translate such changes into weighting coefficients for the AI device. This framework could potentially be applied to many human–AI problems.

We hope the readers will find the articles in this special issue interesting and useful. Finally, we would like to thank all the authors who contributed to this special issue, the reviewers for dedicating their time and providing constructive feedback to the submitted papers, and the editorial staff of *Brain Sciences* for their support.

Conflicts of Interest: The authors declare no conflict of interest.

References

1. Lebedev, M.A.; Opris, I.; Casanova, M.F. Augmentation of Brain Function: Facts, Fiction and Controversy. *Front. Syst. Neurosci.* **2018**, *12*, 45. [CrossRef] [PubMed]
2. Cinel, C.; Valeriani, D.; Poli, R. Neurotechnologies for Human Cognitive Augmentation: Current State of the Art and Future Prospects. *Front. Hum. Neurosci.* **2019**, *13*, 13. [CrossRef]
3. Ayaz, H.; Dehais, F. *Neuroergonomics: The Brain at Work and in Everyday Life*, 1st ed.; Academic Press: Cambridge, MA, USA, 2019.
4. Ruf, S.P.; Fallgatter, A.J.; Plewnia, C. Augmentation of working memory training by transcranial direct current stimulation (tDCS). *Sci. Rep.* **2018**, *7*, 876. [CrossRef] [PubMed]
5. Valeriani, D.; Matran-Fernandez, A. Past and Future of Multi-Mind Brain-Computer Interfaces. In *Brain-Computer Interfaces Handbook: Technological and Theoretical Advances*, 1st ed.; Nam, C., Nijholt, A., Lotte, F., Eds.; CRC Press: Boca Raton, FL, USA, 2018; Volume 1, pp. 685–700.
6. Farwell, L.A.; Donchin, E. Talking off the top of your head: toward a mental prosthesis utilizing event-related brain potentials. *Electroencephalography Clin. Neurophys.* **1988**, *70*, 510–523. [CrossRef]
7. Rezeika, A.; Benda, M.; Stawicki, P.; Gembler, F.; Saboor, A.; Volosyak, I. Brain–Computer Interface Spellers: A Review. *Brain Sci.* **2018**, *8*, 57. [CrossRef] [PubMed]

8.   Tonin, A.; Birbaumer, N.; Chaudhary, U. A 20-Questions-Based Binary Spelling Interface for Communication Systems. *Brain Sci.* **2018**, *8*, 126. [CrossRef] [PubMed]

9.   Ramele, R.; Villar, A.J.; Santos, J.M. EEG Waveform Analysis of P300 ERP with Applications to Brain Computer Interfaces. *Brain Sci.* **2018**, *8*, 199. [CrossRef] [PubMed]

10. Nayak, T.; Zhang, T.; Mao, Z.; Xu, X.; Zhang, L.; Pack, D.J.; Dong, B.; Huang, Y. Prediction of Human Performance Using Electroencephalography under Different Indoor Room Temperatures. *Brain Sci.* **2018**, *8*, 74. [CrossRef] [PubMed]

11. Zander, T.O.; Kothe, C. Towards passive brain-computer interfaces: Applying brain-computer interface technology to human-machine systems in general. *J. Neural Eng.* **2011**, *8*, 025005. [CrossRef] [PubMed]

12. Cavazza, M. A Motivational Model of BCI-Controlled Heuristic Search. *Brain Sci.* **2018**, *8*, 166. [CrossRef] [PubMed]

brain
sciences

MDPI

Review

Brain–Computer Interface Spellers: A Review

Aya Rezeika, Mihaly Benda, Piotr Stawicki, Felix Gembler, Abdul Saboor and Ivan Volosyak *

Faculty of Technology and Bionics, Rhine-Waal University of Applied Sciences, 47533 Kleve, Germany;
aya.rezeika@hochschule-rhein-waal.de (A.R.); mihaly.benda@hochschule-rhein-waal.de (M.B.);
piotr.stawicki@hochschule-rhein-waal.de (P.S.); felix.gembler@hochschule-rhein-waal.de (F.G.);
abdul.saboor@hochschule-rhein-waal.de (A.S.)
* Correspondence: ivan.volosyak@hochschule-rhein-waal.de; Tel.: +49-2821-8067-3643

Received: 21 February 2018; Accepted: 27 March 2018; Published: 30 March 2018

Abstract: A Brain–Computer Interface (BCI) provides a novel non-muscular communication method via brain signals. A BCI-speller can be considered as one of the first published BCI applications and has opened the gate for many advances in the field. Although many BCI-spellers have been developed during the last few decades, to our knowledge, no reviews have described the different spellers proposed and studied in this vital field. The presented speller systems are categorized according to major BCI paradigms: P300, steady-state visual evoked potential (SSVEP), and motor imagery (MI). Different BCI paradigms require specific electroencephalogram (EEG) signal features and lead to the development of appropriate Graphical User Interfaces (GUIs). The purpose of this review is to consolidate the most successful BCI-spellers published since 2010, while mentioning some other older systems which were built explicitly for spelling purposes. We aim to assist researchers and concerned individuals in the field by illustrating the highlights of different spellers and presenting them in one review. It is almost impossible to carry out an objective comparison between different spellers, as each has its variables, parameters, and conditions. However, the gathered information and the provided taxonomy about different BCI-spellers can be helpful, as it could identify suitable systems for first-hand users, as well as opportunities of development and learning from previous studies for BCI researchers.

Keywords: Brain-Computer Interface (BCI); Speller; Graphical User Interface (GUI); Steady State Visual Evoked Potential (SSVEP); P300; Motor Imagery (MI); hybrid

1. Introduction

In this review, we primarily focus on the recent advances in the field of Brain–Computer Interface (BCI) spellers for different electroencephalogram (EEG) signals' features. These speller systems are usually a graphical representation of letters, numbers, and symbols which are controlled using different BCI types for spelling and typing. Audio output can also be included by modern speech synthesis/voice recognition systems.

The majority of research papers in the BCI field focus mainly on the development of the system's back-end to improve the signal processing algorithms and boost the performance of the system (see the Research Methodology section). Our assumption is that, as the Graphical User Interface (GUI) of the BCI speller is the front-end, it is the first parameter which the end-user would judge on a BCI-speller, and, therefore, more attention should be given to it.

The goal of this paper is to describe and gather details about some unique and successful BCI-speller systems (from our point of view), specifically those published during this decade. The older state-of-the-art systems are discussed in this review as they are very well known and represent the basis on which many of the newer developments were built on. These systems were mainly developed with the objective of creating possible communication methods for patients suffering from motor neuron disease (MND) or with the goal of providing an initial proof of concept through examining the reliability of such systems by testing them with healthy subjects.

First of all, this review will benefit many researchers in the field as it provides a reference point giving an overview of the most successful BCI spelling systems. Consequently, it makes it easier and faster to go through many studies, facilitating the initial phase of a new research. Additionally, this review lists the improvements of different BCI-spellers, while highlighting the recent improvements and changes made with respect to the past and also presenting a taxonomy and classification of different features of such systems. This uncovers new development opportunities for further studies and underlines the relevant know-how for fresh researchers in the field.

Another group which could benefit from this paper are the MND patients, their families, and caregivers, as they are the main targeted end-users for BCI spelling applications. As nowadays internet research is a common skill, many patients' family members conduct internet lookups to find suitable rehabilitation systems or any new technology, which might help their afflicted relatives. Finding a review paper listing different options and developments of BCIs for communication (spelling) might be beneficial. It is difficult for healthy users and designers to anticipate the needs of an afflicted person. This review might encourage more patients to be willing to contribute with their opinions and testing for such systems. A BCI speller is characterized by features which attract the end-user. These characteristics are presented in an "easy-to-understand" taxonomy chart. End-users with basic or no prior understanding of the field could build a general knowledge about BCI spellers by reading through this review. The expectation is that there are potential end-users who are interested to learn about BCI. This review offers a smooth start from which MND patients (and patients with similar symptoms) with no previous knowledge or naïve understanding of BCI could begin improving their quality of life by using such systems for communication purposes. Publishing this review as an open access article might also reach more potential users and introduce them to BCI for the first time.

The review is structured as follows: Section 2 "Brain–Computer Interface" is a brief introduction and explanation of BCI in general, focusing on the relevant concepts in this review. Section 3, "Research Methodology", describes the construction methodology of this review. Section 4, "Review of BCI Spellers", presents the different types of BCI-speller GUIs, showing their features and characteristics with a short discussion concerning the described systems in each subsection. In Sections 5 and 6, "Discussion" and "Conclusions", a general discussion about BCI and similar systems is presented, while expressing our personal opinion about the upcoming development opportunities in the field.

2. Brain–Computer Interface

MNDs affect how the brain communicates with the other organs in the body by disrupting neurological networks; they mostly affect the motor control of the muscles. They include Amyotrophic Lateral Sclerosis (ALS). Similar symptoms are shown for Locked-in Syndrome (LIS), brainstem stroke, brain or spinal cord injury, cerebral palsy, muscular dystrophies, and multiple sclerosis, which eventually cause the afflicted patients to lose their ability to control voluntary muscles, mostly consisting of the skeletal muscles and the tongue, thus causing functional and cognitive disabilities. More details on MNDs can be found in [1]. As a result, these patients find it increasingly difficult to communicate with their surroundings, as they cannot speak or even use their hands for sign language.

To help patients to regain their social life, an alternative way of communication is needed. An example of such communication systems, which has been around for years, are the eye-tracking spelling systems, which depend on the movement of the eye that controls a cursor on a virtual keyboard and selects the desired letters [2]. Also, a simple eye blinking can be used as a communication method. Such and similar systems might not be suitable for some patients who have lost the ability to precisely control fine ocular movements or who experience uncontrollable head movements [1,3].

A solution which would allow these patients to communicate is the utilization of modern BCI. A BCI system allows people to communicate through brain signals without the need of any muscular movement. It provides an artificial output that is different from the usual natural output of the nervous system, which is disabled in most MND patients.

There are many methods to monitor the brain's activity. One of the most common methods for measuring brain waves, and our focus during this review, is the electroencephalogram (EEG) [4]. EEG is a non-invasive measurement technique widely used in almost all modern BCI applications, more practical than Electrocorticography (ECoG), which requires an opening through the skull to directly access the brain tissue [4]. The main reasons why EEG is so common are: EEG equipment is relatively inexpensive, portable, simple to set-up, and provides a signal with high time resolution compared to other non-invasive methods for monitoring brain activity like Magnetic Resonance Imaging (MRI) or Positron Emission Tomography (PET), just to mention a few [5]. Also, non-invasive BCIs could become a useful tool to be utilized and tested by healthy individuals for research and development of applications [4].

After measuring and recording the brain activity in a BCI system, specific features of the signal are extracted and analyzed by the computer. This output has the potential to serve as a BCI application which might replace, restore, enhance, supplement, or improve the function of the central nervous system [6].

Over the recent years, BCI researchers have been developing various applications which might be useful for MND patients in particular. One of the most commonly studied applications is the BCI-speller. Usually, a BCI-spelling application allows the users to communicate with the environment using a GUI. The GUI displays letters, numbers, and special characters. With the aid of the brain signal recorded and analyzed by the BCI system, the user selects the desired character and types it on the screen or other output displays. Farwell and Donchin presented the first spelling application in 1988 [7]. Promising accuracy levels and typing speeds have been presented in the literature since then. Consequently, BCI spelling applications were further developed, allowing people to communicate directly through the measurements and direct interpretation of brain activities. In general, the loss of communication for such patients affects their quality of life negatively as presented in [8]. Subjects using a BCI-speller can be more independent and can even regain their social life to a relatively high extent.

The measured brain activity from the BCI is interpreted with the intention of selecting the desired key (letter, number, or symbol) shown on the screen. In contrast to standard physical keyboards used traditionally in most computer systems, where the user selects the desired key by physically pressing it, in a BCI system, the user selects a key by looking at it (or by other sensory modalities in some cases), and the letter will be "pressed" by the computer according to the measured and classified brain signals.

The performance of BCI-spellers is commonly measured by calculating the accuracy and the Information Transfer Rate (ITR) of the system. The accuracy is calculated by dividing the number of correct commands by the total number of commands. The commonly used ITR was introduced by Wolpaw in [9], originally presented much earlier, as discussed in [10]. The ITR combines the accuracy and the system's speed in one variable and it is expressed as the number of error-free bits per time unit. It is important to note that the ITR may be calculated in different ways (e.g., on the level of commands or of the letters) in different types of BCIs. It can only be used objectively to compare the performances of systems of the same type.

This review focuses on the main EEG paradigms used by the vast majority of BCI spellers: Event-Related Potentials (ERP) (mainly P300 and Steady-State Evoked Potential (SSEP)) and motor imagery (MI, also called Event-Related Desynchronization/Synchronization (ERD/ERS)) [4,5].

2.1. Event-Related Potential (ERP)

ERPs are electrocortical signals which can be detected and measured using EEG, during or after a sensory, motor, or psychological event. They usually have a fixed known time delay to a stimulus and a different amplitude compared to the spontaneous EEG activity. ERPs are less frequent and more localized than the normal EEG-measured signal. Different ERPs can be evoked using different types of stimulus (events), and the evoked ERP is characterized by a specific time delay and/or location where it was generated. The two most common ERPs are the P300 and the Steady-State Visual Evoked Potential [11].

The P300 wave is a type of event-related potential which occurs in the human brain as a positive deflection with a time delay of around 300 ms after a specific event has occurred [12] (although the timings may vary, as discussed in [13]). The P300 signal is usually intensified over the central parietal region of the brain and can be detected using EEG. The event which stimulates the P300 is known as "the oddball paradigm" [7]. Accordingly, this paradigm consists of three main prerequisites [6]:

- A subject is presented with a series of stimuli or events; each of them belongs to one of two classes (e.g., a desired or an undesired event)
- One of the classes is less frequently presented than the other class (a rare event versus a usual event)
- The subject needs to pay attention to one of the stimuli when it occurs (e.g., counting how many times a particular letter will flash, which is the rare event).

The rare events induce the P300 signal in the brain. Researchers have developed both visual and auditory stimuli to induce a P300 signal for different systems and applications (Visual Evoked Potential (VEP) and Auditory Evoked Potential (AEP)). One of the first BCIs using the P300 signal is a speller developed by Farwell and Donchin in 1988 (Figure 1a) [7], which used visual stimulation for the "the oddball paradigm". Hill et al. in 2005 [14] introduced the first P300 BCI based on auditory stimuli. In 2014, a novel auditory speller, named "charstreamer", was presented by Höhne et al. [15]. As for other types of sensory stimulation, in [16], a tactile P300 BCI was developed by fixing vibration motors at different locations around the participant's waist. The user had to focus on the vibration at the desired location and ignore all the others to elicit a P300 signal. More recently, researchers started experimenting with placing motors on different parts of the body, such as the back or the hand of the user, with the aim to improve the tactile P300 BCI performance [16,17].

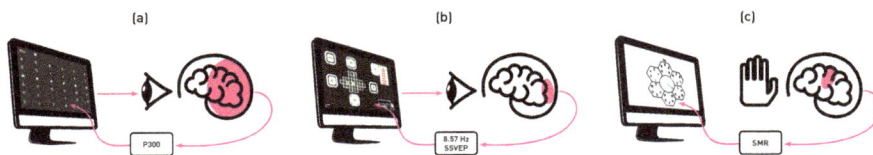

Figure 1. Schematic representation of three major Brain–Computer Interface (BCI) paradigms: (**a**) P300 paradigm. The oddball paradigm causes a P300 signal in the brain of the user which is then interpreted by the BCI system, resulting in the selection of the desired letter; (**b**) Steady-State Visual Evoked Potential (SSVEP) paradigm. Five different frequencies are shown on the screen in this example, as discussed later. When the user gazes at one of them, an SSVEP signal with the same frequency (as well as its harmonics) is elicited in the visual cortex of the brain. The measured electroencephalogram (EEG) data are analyzed by the BCI, and a command is sent to the computer to select the target; (**c**) Motor imagery (MI) paradigm (with a schematic representation of a Hex-O-Spell application, as discussed later). The imagination of the movement of limbs (in this picture an imaginary movement of an arm) induces a sensorimotor rhythm (SMR) signal which is detected and analyzed by the BCI system, and a feedback is sent to the computer to control the movement of the green arrow for letter selection. In this case, the presence of an external stimulus is not required.

The Steady-State Evoked Potential (SSEP), specifically the Steady-State Visual Evoked Potential (a type of VEP) (Figure 1b), is characterized by positive and negative fluctuations in the EEG signal which are responses to a visual stimulus. For example, light is flashing, an image is appearing/disappearing, or a pattern is presented with a certain frequency. SSVEP is recognizable in the EEG recordings as voltage oscillations which are further processed to detect their unique features, such as frequency and amplitude. When the external visual stimulus is flickering at a specific constant frequency, an SSVEP is elicited with a peak frequency matching the stimulus (as well as its harmonics), mainly in the visual cortex, located in the occipital region of the brain, given that the subject's eyes are fixated on the stimulus. Usually, a frequency analysis technique, such as Fast Fourier Transform (FFT), is used to detect the stimulation frequency [6].

In a standard SSVEP system, taking a spelling application as an example, the targets can be individual letters or groups of characters or command boxes. Each target flickers with a unique frequency. This is also known as frequency-modulated Visual Evoked Potential (f-VEP). Another well-known type of VEP is the so-called code-modulated Visual Evoked Potential (c-VEP). Instead of using a constant flickering frequency, the stimulus is a pseudorandom swapping of orthogonal patterns [18]. It is worth adding that tactile stimuli were also used to elicit an SSEP response in a BCI system [19].

Another type of VEP is the Motion-Onset Visual Evoked Potential (mVEP). The above mentioned VEPs depend mainly on light flashes or patterns. In 2008, Fei Guo et al. [20] presented a different approach, the first BCI system based on mVEP. In [20], visual responses from the dorsal pathway of the visual system were utilized, which led to the use of more elegant visual stimuli. The mVEP paradigm has been used for several years to investigate human brain motion processing [21]. It is typically comprised of three main peaks: P1, N2, and P2. The N2 peak, with a latency of 160–200 ms, is predominantly motion-specific, and the P2, with a delay of about 240 ms, is elicited with more complex visual moving stimuli [22–24]. The mVEPs are usually elicited by a pre-defined simple motion of the visual targets.

2.2. Movement Imagination

The sensorimotor rhythm (SMR) (Figure 1c) can be recorded over the motor cortex with the contribution of some somatosensory areas. During movement, Motor Imagery (MI) and movement preparation the SMR can be decreased or increased; these options are known as Event-Related Desynchronization (ERD) and Event-Related Synchronization (ERS), respectively [11]. During ERD, the signal drifts and becomes lower than a specific baseline, which might be due to the desynchronization of the activities of specific areas of the brain [11]. On the other hand, during ERS, the signal measured during movement is stronger when compared to a baseline. The signal location varies depending on which limb is moving and on which side of the body the specific movement is taking place. It was also discovered that the imagination of a movement without actually performing it elicits a similar EEG signal [25]. Even though this signal is weak in comparison to ERP and VEP, leg and arm movements can be distinguished, as well as the side of the upper limb (left or right) [6].

3. Research Methodology

Literature research was conducted according to the PRISMA guidelines [26] (PRISMA diagram shown in Figure 2), using the IEEEXplore Digital Library (incl. conference proceedings) and further online databases through Web of Science (WOS). In both, the search was conducted using the search terms "BCI" AND "speller", and the dates were restricted from 2010 to January 2018. First, the search was done without constraining the years, as a test. WOS showed 316 results, and IEEE showed 213 results, giving a total of 529 (including duplicates if any). Later, the search was performed with time restriction. WOS showed 287 results, and IEEE showed 173 results, giving a total of 460. From this observation, we deduced that the steep growth of research during this decade deserved a deeper look, taking into account the origin of these developments from earlier years.

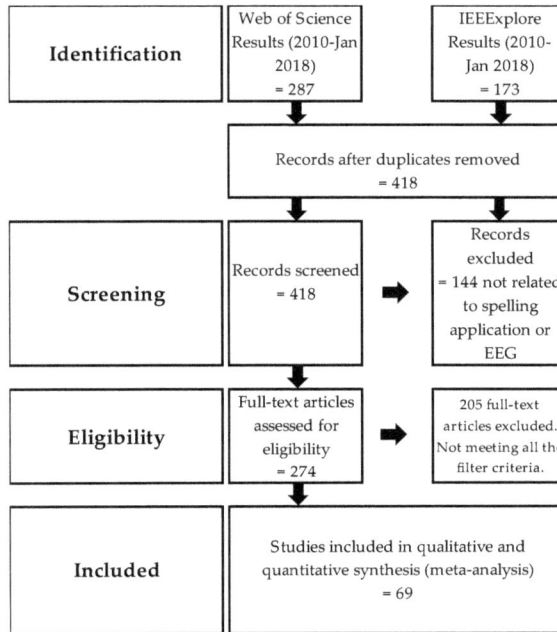

Figure 2. PRISMA chart.

Both lists (WOS time-restricted and IEEE time-restricted) were extracted for analysis. Duplicates were detected, as some of the IEEE conference proceedings and journal papers were listed on the WOS database. First, the duplicates were removed, resulting in a total of 412 papers and articles. Then, the papers were checked and classified manually according to BCI type, output type, number of subjects, type of subjects, and purpose of the research. The final step was to determine which papers were relevant to the topic of our review. The filtration of 412 remarkable papers and articles according to the below-mentioned classification criteria used in this review was a delicate process, which took an extensive number of working hours:

- Non-invasive BCI = EEG-based BCIs
- Only visual stimuli or movement imagery
- The purpose of the research or the aim of the research is the development of a new Graphical User Interface (GUI) for a BCI speller system OR a clear modification of an existing GUI
- Published between 2010 and January 2018 (with few exceptions).

BCI Spellers Taxonomy

All BCI speller systems can also be categorized according to the following characteristics: dependent or independent, synchronous or asynchronous, with regard to the stimulus type and gaze dependency.

The P300 and SSVEP depend on visual stimulation to induce a specific brain activity which can be later interpreted by the BCI system. Thus, a stimulus must be physically present in the environment to initiate the required signal. The P300 and VEP require a structured environment to present external stimuli. Such BCIs are usually implemented as dependent BCI. The MI-BCI, however, depends on the imagination of the movement of any limb, whole-body activities, performing of specific cognitive tasks, relaxation, etc. This imagination initiates a brain activity in the motor cortex region of the brain, which can be detected and interpreted by the BCI system. In this case, only the subject is responsible for creating and firing up the desired brain signal; MI-BCI systems are therefore classified as independent BCIs, as no external (e.g., visual) stimuli are required.

A synchronous system limits the time intervals when the BCI will process the measured and analyzed brain activity into actions [27]. It provides a starting point and measures a specific brain signal which happens afterward. Thus, the system limits and specifies the time at which the BCI can use the measured activity to produce a useful output. In the meantime, the protocol might provide the user with cues to alert the user to get ready and prepare for the coming stimulation phase. Usually, synchronous BCIs do not consider the possibility that, at a specific point in time, the user has no intent to use the system. The commonly used P300 speller application based on [7] is a typical dependent synchronous BCI. The P300 stimuli occur for a specific, pre-defined period in which the user has to focus the attention on the displayed GUI for a meaningful output. In addition, some MI applications specify the time slot where the user should imagine the movement, i.e., the user has to wait for a cue to perform a movement imagination, otherwise an error would occur [6].

The asynchronous (or self-paced) protocol is simply the opposite. The user has the ultimate control over the system, whenever they desire. Asynchronous BCIs result in a more natural and dynamic interaction between the user and the system. The user does not have to wait for a cue to control the system. The SSVEP-based BCI system can also be created as an asynchronous BCI, e.g., the user directs his/her focus of attention on the flickering SSVEP stimuli (when the user shifts her/his gaze away from the stimuli = no classification) [6].

Moreover, recent studies are investigating BCI-spellers according to the type of attention needed, whether it is overt or covert. Overt attention occurs when eye movement is involved in paying attention to a specific visual space or region. Covert attention is more of a mental attention and not a specific visual attention. While the eyes are fixed, the attention is shifted mentally to the desired focus point; this attention is not directly associated with eye movements. These two types of attention allow another categorization of BCI spellers: gaze dependency (gaze-independent versus gaze-dependent). Many studies are seeking to achieve a gaze-independent speller that requires minimum ocular muscles movement. However, gaze-dependent spellers are most common.

Another characteristic of a BCI speller is the type of stimulus presented to the user. Most SSVEP systems rely on flickering stimuli with constant frequencies (each stimulus has its own unique frequency). P300 spellers, as mentioned before, are based on the oddball paradigm according to which characters are flashed periodically in a predefined order (pseudorandom). A number of them apply the oddball differently by animations, flashing faces, or movement.

Table 1 shows the taxonomy for the different BCI paradigms according to this categorization for the studies listed in this review. The filtered data resulted in 69 papers which meet all the filtration criteria. Of these papers, 45 are based on P300, 16 on SSVEP, and 4 each are based on MI and hybrid BCI. At the top of Table 1, the contribution percentages, from the total number of PRISMA results, of each BCI type are presented. The figure also categories the 69 systems according to the above-described taxonomy. This table could assist our readers to select the speller which falls into the category of interest or even find the appropriate speller on the basis of other characteristics. It also highlights some development opportunities, for example, none of the 45 P300-based spellers is asynchronous. It also underlines the features of each speller described in this review.

Table 1. BCI spellers' taxonomy. The table also classifies the papers presented in this review according to the suggested taxonomy.

BCI Paradigm		P300 45 Studies 65% of Total	SSVEP 16 Studies 23% of Total	MI 4 Studies 6% of Total	Hybrid 4 Studies 6% of Total
Operation Modality	Asynchronous 21.7%	0.0%	68.8% [28–38]	75.0% [39–41]	25.0% [42]
	Synchronous 78.3%	100.0% [7,43–88]	31.3% [89–93]	25.0% [94]	75.0% [95–97]
Gaze Dependency	Gaze Independent 15.9%	15.6% [74,78–80,82–84]	6.3% [37]	75.0% [39,40,94]	0.0%
	Gaze Dependent 84.1%	84.4% [7,43–54,56–60,62–73, 75–77,81,85–88]	93.8% [28–36,38,89–93]	25.0% [41]	100.0% [42,95–97]
Selection Modality	Direct Target Selection 92.8%	100.0% [7,43–88]	87.5% [29–36,38,89–93,98]	25.0% [94]	100.0% [42,95–97]
	Moving Cursor 7.2%	0.0%	12.5% [28,37]	75.0% [3–41]	0.0%
Stimuli Modality	Constant Flashing 26.1%	0.0%	100.0% [28–38,89–93]	0.0%	50.0% [95,96]
	Periodic Flashing 58.0%	82.2% [7,43,44,46–49,57–60, 62–81,83–88]	0.0%	0.0%	75.0% [42,95,97]
	Moving/Animation 13.0%	17.8% [45,50–54,56,82]	6.3% [93]	0.0%	0.0%
	No visual Stimuli 5.8%	0.0%	0.0%	100.0% [39–41,94]	0.0%
Word Prediction	Yes 14.5%	13.3% [62–65,75,76]	18.8% [28,37,38]	25.0% [94]	0.0%
	No 85.5%	86.7% [7,43–60,66–74,77–88]	81.3% [29–36,89–93]	75.0% [39–41]	100.0% [42,95–97]

4. Review of BCI Spellers

Many types of BCI spellers have been developed over the years. This review primarily discusses the work done since the beginning of this decade considering the development of novel Graphical User Interfaces (GUI) of BCI spellers or improvements on the already existing and widely known GUIs.

As presented in the previous Section 3, over 400 publications were issued since 2010, with the aim of developing BCI-spellers. The PRISMA guidelines analysis showed that only ~18% of these studies directly targeted the improvement of the GUI design. Although other developmental aspects of a BCI system are very important to achieve a high-performing BCI-speller, the GUI is the first thing the end-user would encounter when dealing with such systems and it very often gets the least attention in the development process. In our opinion, the user-friendliness and the performance of the system are important factors. In addition, the design of the GUI might directly affect the performance parameters (accuracy and ITR).

In total, 75 relevant papers are discussed in this section of the review. The section classifies the spellers according to the type of BCI system used. This classification was presented for two main reasons: (1) Different types of BCIs might perform differently for the same user. The end-user might be interested in reading about a specific type of BCI, if from a previous experience he/she knows that this is the most suitable for him/her.; (2) Usually, each research team is working on a specific type of BCI paradigm. Categorizing these papers in this manner would also be beneficial for the readers.

4.1. P300 Spellers Based on the Matrix Speller

The first P300-based speller was introduced by Farwell and Donchin [7], and Figure 3a is showing a similar design to their GUI. It was the first BCI application based on P300. It consisted of a 6 × 6 matrix of flashing symbols displayed on a monitor. The items were organized in rows and columns (row–column paradigm, RCP), which were intensified in a random order, constituting an "oddball" paradigm. As this matrix consisted of six rows and six columns, at least 12 flashes were needed to flash each column and row once. The subject focused his/her attention on the target letter and was asked to count the number of flashes to help focus. The flashing of the row and the column which contained the desired target would produce a P300 wave in the EEG signals. The EEG signal was then processed, and the P300 signal was correlated to the order of occurrence of the flashing of the presented rows and columns. The analysis of these data resulted in the exact row and column which induced the P300 signal, the intersection of which was the selected letter.

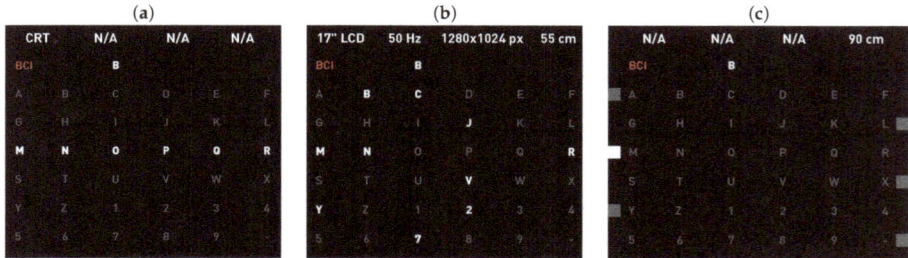

Figure 3. Graphical User Interface (GUI) of a modern P300 speller: (**a**) Matrix Speller inspired by the matrix developed by Farwell and Donchin in 1988 [7], shown during the intensification of the third row. (**b**) The random intensification similar to the one discussed in Yeom et al., 2014 [43]. (**c**) A view of the Edge Paradigm from Obeidat et al., 2015 [44] showing the intensification of the edge point next to the third row. All the above figures show "BCI" as the target word during spelling and "B" as an already selected character. Figures modified from the cited sources.

The maximum accuracy reached in this study was 95% at a speed of 12 bits/min. This means a character can be selected from the matrix in approx. 26 s. This can be considered as very slow compared to conventional typing systems for healthy people; however, it can mean a lot for a person with no other means of communication.

The Matrix Speller is the base of most P300 BCIs. Researchers conducted many developments to make it faster, to achieve better classification, accuracy, and user-friendliness. The first research conducted by Farwell and Donchin had only four healthy subjects; however, over the years, many subjects (healthy and with different disabilities) have been testing their concept.

Farwell and Donchin proved the concept that P300 can be used for selecting a specific choice using the special arrangements of characters in the matrix, confirming that the P300 can be used for a communication application.

4.1.1. Stimuli Variations

Many variations were proposed based on the GUI of the P300 Matrix Speller. One of the main variations is the change of the flashing stimuli. In 2010, Liu et al. [45] tested and discussed different types of intensification techniques for the Matrix Speller. Instead of just flashing individual symbols or rows and columns, as the flashing can be uncomfortable for some subjects, they used graphical effects like translations, rotations, zoom in/out, pattern rotation, and sharpening types. This stimulation technique can be applied to bigger menus with the advantage of a lower number of flashes, for a faster system. The different stimulation techniques suggested were a relative success. As a result, the best intensification was not the same for all subjects. This means the speller can be personalized individually for the best performance of each subject. Some types showed better results than typical flashing or a simple color change.

In [46], the 6 × 6 matrix speller was divided into four 3 × 3 submatrices. Randomly, the character was flashed from each submatrix once, so that, in total, only nine trials were produced. Another form of a submatrix stimulation was discussed by Eom et al. [47], called Sub-Block paradigm. Only a 2 × 3 submatrix of the 6 × 6 matrix speller was highlighted and not the entire row/column sequence. Further research showed that the change of the flashing patterns for individual characters was possible.

In [48], only 7 or 9 flashes per trial were required, compared to the original matrix speller which required 12 flashes (one for each row and column), making the application faster. The nine flashes showed the highest accuracy and corresponding ITR, that were 92.9% and 14.8 bits/min, respectively, while the 12 flashes showed 88.0% accuracy and 10.1 bits/min ITR and the seven flashes showed 68.8% accuracy and 5.3 bits/min for ITR. The highest ITR reached was 17.3 bits/min, but with slightly lower accuracy. The aim was to achieve faster spelling speed and minimize the errors. A similar approach was shown by Polprasert et al. [49]. Similarly, the Random Set Presentation (RSP) was studied and tested by Yeom et al. in [43] to show the effect of a random intensification of characters, by flashing the characters in a random order (Figure 3b).

Fazel-Rezai in [99] discussed the "adjacency problem". Flashes next to the target seemed to be distracting the user and sometimes resulted in the wrong feedback as well as in the increasing of the problem of crowding, which refers to the difficulties in identifying a target if many similar objects surround it. In [43,46,47], the main aim was to avoid the adjacency-distraction effect and double-flashing errors. The system mentioned in [46] showed a higher performance than [47] with a mean accuracy of 99.70% and ITR of 26.8 bits/min. Dividing the 6×6 matrix into smaller matrices can be more comfortable for the users' eyes, especially when only one character per submatrix was flashed at a time as discussed in [46]. In addition, in [43], the adjacency-distraction error was avoided by random-set representation, and, when flashing single characters randomly, no two adjacent letters were intensified at the same time.

The edges paradigm (EP) was introduced by Obeidat et al. [44] to overcome the mentioned challenges. The difference between the EP and the RCP presented in the Matrix Speller were the flickering points, which were added to the left of each odd row, to the right of the even rows, below the odd columns, and at the top of the even ones; the first step (row selection) is shown in Figure 3c. These points were intensified by increasing the illumination rather than by normal flashing, and the characters were fixed. For the selection of the desired letter, the subject first needed to focus on the edge of the row which contained the target letter. Then, during the second stage, the subject needed to focus on the edge point corresponding to the column which contained the target.

The edges paradigm was one of the most successful paradigms for solving the adjacency problem. As only the edges of the rows and columns were flashing, and not the characters, the flashing of characters was avoided, thus solving the adjacency problem and the double-flashing problem and reducing the discomfort which might result from an extended use of a flashing RCP. Although the mean ITR of the system was not as high as that of other presented systems, it still showed high accuracy. A total of 14 participants answered a questionnaire rating the levels of fatigue and comfort comparing the RCP and the EP. The results reported that the EP caused less fatigue and was more comfortable to use than the RCP. The advantages of this system were notable, and, as for the relatively low ITR, it can be adjusted by training, for example.

4.1.2. Familiar Faces and Symbols

Numerous studies in the field of human face processing have revealed that the visual perception of familiar faces strongly involves several ERPs, which may be exploited for improving the classification. In particular, using faces well known to everybody in a given culture should lead to high and relatively stable effects across individuals. In [50], a 6×6 Matrix Speller was described; however, each character was superimposed by a semitransparent picture of a familiar (famous) face. In this study, they used faces of Albert Einstein or Ernesto 'Che' Guevara. The characters were intensified by the appearance of the familiar face behind the stimulated row or column. The paradigm was compared with a classic Matrix Speller where the new familiar face paradigm showed faster target selection and comparably high accuracy due to the fact that the familiar faces induced a higher ERP response.

The prototype in [51] was further studied in [52] by modifying the familiar face color to green. The green colored faces showed even a higher ERP response. A similar spelling system was used in [53] as a two-stimuli spelling system, utilizing familiar face and character flashing to increase the speed of spelling. The speller proved to be two times faster than the classical Matrix Speller. In [54], a similar approach was studied. A classical row/column paradigm and a random stimulus presentation of the row/column paradigm were compared to two proposed paradigms. The first paradigm presented random flashing of a self-face picture, while the second paradigm presented random flashing of non-self-face pictures. Another similar study was done more recently in [55].

Almost all of the mentioned spellers which use familiar faces showed a relatively high performance, higher accuracy, and faster ITR. The highest average ITR in this topic was ~80 bits/min reported in [53] with an accuracy of 81.25%. The highest mean ITR was reported in [54] with a 90.7% accuracy (more details, also about the other studies, are shown in the summary tables in the Discussion Section). The goal of using familiar faces was to have more effective visual stimuli which would elicit a stronger ERP signal. This would result in a more accurate classification. Also, combining the familiar face stimulus with another type of stimulus, like random-set-representation, was a promising development. It combined both systems' advantages, avoiding main problems like adjacency-distraction and double-flashing errors.

In [56] a study by Kathner et al., rows and columns in a 5 × 5 matrix were flashed with the display of a yellow smiley face in a Virtual Reality (VR) environment. Using VR headset, two screens were tested: a full-view screen where the user could see the whole matrix and a second screen where the user could only see the part of the matrix on which he/she was focusing on, and head movement was required to visualize the rest of the matrix. It was tested on a patient with LIS, who showed adequate control over the BCI system. The paradigm combined with the Virtual Reality resulted in a fast and accurate BCI speller system. The system addressed one of the most challenging problems in the BCI field, i.e., portability. Using a virtual reality headset as a display eliminated the use of big computer monitors or other screens. The relatively decent performance was reported (see the summary tables in the Discussion Section). This BCI speller was based on a stimulus different from flashing characters; yellow smiley faces appeared over the characters for intensification. The mentioned stimulus type was similar to the familiar faces stimulation, resulting in a stronger ERP signal. The system performance was tested against the performance of the same interface on a 22″ monitor, showing no significant differences. However, the second view proposed in the system, where the user had to move his/her head to see the rest of the matrix, can be impractical for some neuromuscular disease patients.

4.1.3. Variation of Letters Arrangement

Letters arrangement was tested as another parameter of the matrix speller. In [57], the arrangements of letters were changed according to the feedback from a built-in dictionary, which arranged the letters according to their usage frequencies. The more the user used a letter, the more accessible was the position in which it was placed.

In [58], the letters were arranged in a 7 × 7 matrix according to the frequency of their usage in the English language. Interestingly, for this system, the speller was tested with ten neuromuscular disease patients, plus ten healthy subjects. It showed higher accuracy than a normally ordered ABC interface, assuming less workload, but a lower ITR. From the questionnaires, most participants preferred the proposed interface to the ABC. The patients' results appeared to have higher accuracy in the tested interface than in the ABC interface. The healthy participants had higher ITR in the ABC interface; however, ITR was not significantly different for afflicted subjects in the ABC interface and frequency-based interface.

Additionally, in Jin et al. [59], a laptop-keyboard-like matrix was suggested, where a 7 × 12 matrix was presented, in which the letters were arranged according to alphabetical order. The performance was remarkable (ITR 27.1 bits/min, accuracy 94.8% for 21 flash patterns); however, it was only tested on healthy subjects and it comprises many elements which might confuse some users who are not familiar with keyboards.

Inter-character distances were discussed by Sakai and Yagi [60]. The inter-character spaces in a matrix speller were changed and tested from 10 mm, to 25 mm, and to 40 mm. The results showed that the smaller the inter-character spaces were, the higher the P300 signal was. However, smaller inter-character spaces resulted in lower performance, as it was harder to classify the specific letter which caused the P300 response.

4.1.4. Matrix Speller with Predictions

Another investigated approach to make the Matrix Speller faster and more efficient was the addition of a module to the speller which predicted and displayed suggested words for the user to select (for more information about integrating language models into BCIs, please refer to [61]). In 2011, Ryan et al. [62] developed an 8 × 9 matrix which displayed characters, numbers, and other commands on the screen, with the ability to predict the desired word and print suggestions from which the user could choose. A year after, Kaufmann et al. [63] provided a 6 × 6 matrix with predicted words and presented them as an extension of the matrix. The suggested words flashed within the matrix once they were predicted. The user could select the desired word in the same way a target letter would be selected. Later on, a modified matrix speller was added by Akram et al. [64,65], which included a built-in dictionary that displayed suggested words on the side after the user selected a few characters. Each word had a corresponding number. The second step of selection was via a 3 × 3 number matrix, where each number corresponded to one of the suggested words.

The study in [62] was aiming for high ITR without affecting accuracy by adding prediction words. However, the authors suspected that adding a predictive module to the speller might be more cognitively demanding for users because of multi-tasking. This study showed lower accuracy than the usual matrix speller paradigm, but higher ITR. It was indirectly deduced that a predictive system increases the workload on users which can decrease the P300 signal amplitude. This can be enhanced by the training of both, the users and the predictive dictionary. In [63], the solution to the workload problem accompanied with a predictive speller was proposed by displaying the suggested words as part of the matrix. This resulted in less workload on the BCI users, was more comfortable, and also showed a better performance. The system can be modified further to recognize grammar rules and fill in some words for the user. Overall, the spellers with predictions discussed in this review showed promising results, higher ITR than most of the matrix spellers, and a reasonable accuracy (further details in the summary tables in the Discussion Section).

4.1.5. Other Languages

Other P300 speller developments were made by including more interfaces for different languages to allow a broader group of people to benefit from such BCI applications. Developing innovative interfaces for languages based on script was also really important, as it might take a long time to type one word using a BCI speller. Also, most of these non-English interfaces proved to achieve a performance comparable to that of other English spellers.

- Chinese

The Chinese language has a logographic script comprising more than 11,000 characters which are based on strokes. A P300-based BCI has been developed that allows users to input Chinese characters stroke-by-stroke [100]. However, this was not very efficient, as a single Chinese character may consist of 20 or more strokes, and took a long time. Minett et al. [66] showed how a P300 matrix could present an efficient way to type Chinese characters. Also, refs. [67,68] have developed Chinese BCI spellers.

- Arabic

In Kabbara et al. [69], a P300 Matrix Speller was presented in Arabic letters for the first time. A 6 × 5 matrix displays all the Arabic letters in an RCP, where the intensification was the random flashing of rows and columns.

- Korean

In 2011, the first Hangul (Korean script) P300 speller was developed [70]. Hangul has a hierarchical structure entirely different from English; therefore, a two-stage speller was needed. Two-stage speller means that there are two different screens displayed, dividing the symbols.

- Japanese

A conventional Japanese P300-based BCI spelling system consisted of a 6 × 10 matrix. However, Yamamoto et al. [71] proposed a two-phase matrix system; each phase is a 6 × 5 matrix with the option to move between them. This solved the crowding problem as well as decreased the number of flashes needed per trial. A similar approach, another two-phase speller, was tested on ALS patients in [72] and compared to the performance of the conventional system. The two-phase speller was successful in producing higher accuracy rates when tested with ALS patients.

4.1.6. 3D Blocked Matrix Speller

In Noorzadeh et al., 2014 [73] a 3D virtual matrix was suggested. The characters were displayed in 3D blocks instead of the usual 2D screen arrangement. Different flashing techniques were tested for this new design. The study confirmed that this arrangement has the potential to be a more user-friendly GUI.

The speed was proven to be higher in the proposed 3D interface compared to the classic 2D interfaces, since it needed a smaller number of flashes. However, 3D interfaces might need higher computational power. However, still, a 3D design is more attractive and user-friendly.

4.2. Other P300 Interfaces

In this section, other P300 interfaces, which were not a direct development of the original matrix speller, are described and discussed.

4.2.1. Chroma Speller

The Chroma Speller, developed by Acqualagna et al. [74], worked via presenting six differently colored stimuli on a black background, as shown in Figure 4a. A total of 30 characters and symbols were grouped into the six colors for the first selection. When it started operating, the colors flickered in a series manner. The subjects had to focus on the desired color to select it, and the ERP P300 signal was detected and analyzed. After the first selection of a group of characters, the individual characters of the selected group were presented separately on the second screen with row colors similar to the first display, as shown in Figure 4b, with the option to go back to the primary group display if the white box was selected.

The Chroma Speller aimed to achieve a gaze-independent speller system with a minimum workload, as the user had only to focus on the required color (the color which contained the desired character) and not on the individual letter. The proposed system was compared to the Centre Speller (refer to Section 4.2.7.) during the study, showing a higher performance. Consequently, this speller was undoubtedly suitable for patients in advanced stages of ALS, as they face a limited oculomotor control. In addition, the system included an auditory feedback citing the selected letter, which can be helpful for these patients as well. However, on the basis of our knowledge, such system has not been tested yet by ALS patients.

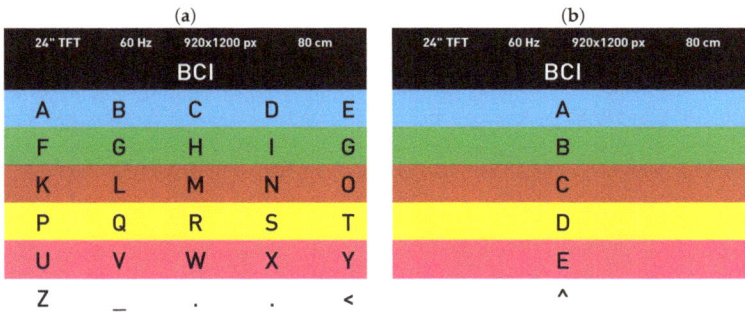

Figure 4. Chroma Speller in its two operating stages [74]. (**a**) The first-stage selection. (**b**) The second stage to select an individual character. The target word here was suggested to be "BCI", and "B" is the target letter. Figures modified from [74].

4.2.2. T9

The first time a T9 (text on nine keys) paradigm was presented as a BCI speller, it was based on auditory stimuli [101]. A modified, visual-based stimuli T9 speller system was introduced in 2015 [75] and is shown in Figure 5, with an integrated dictionary to propose suggested words to save time. T9 is the same approach used in early mobile phones for texting on the number keypad. In this study, they used only eight keys for character input and one as a delete option in case of errors. The user started by typing a few characters of the desired word, then selected one of the suggested words from the same 3 × 3 matrix in the GUI. The targets were highlighted randomly. Figure 5a illustrates how the selection of the letters group occurred, and Figure 5b the second stage, where the selection of a number for the first screen corresponded to one of the suggested words in the list on the right-hand side. This paradigm reduced the typing time significantly, especially compared to other multi-stage spellers. The fast typing speed was credited to the word prediction module embedded in the interface, which was not present in the traditional Matrix speller. Also, as a result of a lower number of stimuli in a T9 interface, the speller was more user-friendly and might have caused less fatigue. As another advantage of having only nine targets, the eye movement was minimized.

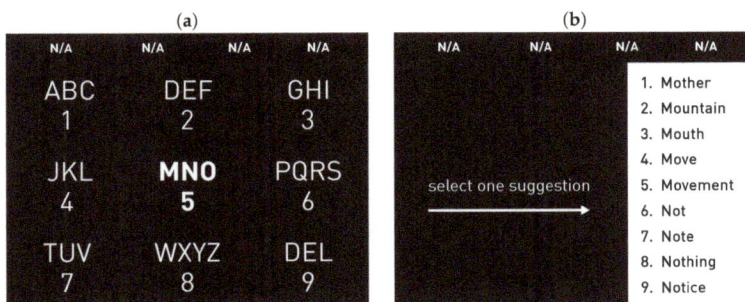

Figure 5. The GUI discussed in [75]. (**a**) The first stage where a target letter was selected; (**b**) Suggested words were displayed with the corresponding number. Figures modified from [75].

A similar T9 system was used in another study in [76] to test its performance with ALS patients and to compare it with a modified Matrix Speller. The T9 showed a faster typing rate with ALS patient compared to the Matrix Speller, revealing a promising performance.

4.2.3. Checkerboard Paradigm

The "Checkerboard" paradigm (CBP) was proposed in [102]. As presented in Figure 6, it was composed of a 9 × 8 matrix of characters and commands, and, to avoid the "adjacency-distraction problem" and the "double flash" issues, the sets of nonadjacent elements were pseudo-randomly flashed [102]. Also, the same paradigm conducted on ALS patients showed higher online accuracy rates for the CBP.

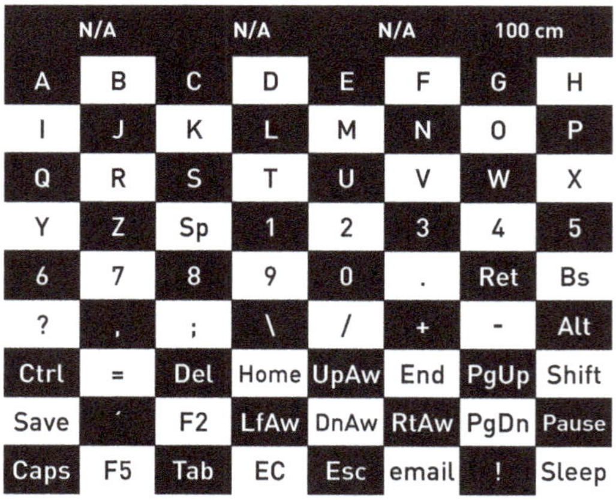

N/A		N/A		N/A		100 cm	
A	B	C	D	E	F	G	H
I	J	K	L	M	N	O	P
Q	R	S	T	U	V	W	X
Y	Z	Sp	1	2	3	4	5
6	7	8	9	0	.	Ret	Bs
?	,	;	\	/	+	-	Alt
Ctrl	=	Del	Home	UpAw	End	PgUp	Shift
Save	'	F2	LfAw	DnAw	RtAw	PgDn	Pause
Caps	F5	Tab	EC	Esc	email	!	Sleep

Figure 6. Checkerboard paradigm similar to the one studied in Townsend et al. [102]. Figure modified from [102].

The half checkerboard paradigm (HCBP) [77] divided the matrix on the screen into two separate regions: "left" and "right". The authors also used electrooculography (EOG) to identify the eye position, so that only the characters in the eye-gaze area would start flashing. This paradigm was targeted to people who can voluntarily gaze at a target and to disabled people who still retain some eye movement. When an area was selected by gazing, it flashed half of the presented 72 targets. The performance of the HCBP was compared with that of the Checkerboard paradigm, resulting in higher accuracy and faster information transfer rates.

The aim of the Checkerboard Paradigm [102] was to compare a new representation of a P300 speller with the RCP. Although the performance of both systems was almost the same, the CBP proved to be less affected by the common errors faced when dealing with RCP. Also, the system was supported by successful trials with ALS patients with better results than with the RCP. Moreover, the participants shared their opinion about the CBP, stating that it was more comfortable and caused less fatigue.

4.2.4. Geospell

We can consider the GeoSpell (Geometrical Speller) as a rearrangement of the P300 Matrix Speller. It was developed by Aloise et al. [78], focusing on covert attention speller. The main concept was to use an N × N matrix, for example, a 6 × 6 matrix, where the total number of characters is N^2 ($6^2 = 36$ characters). Then, the matrix layout was transformed into 2 × N sets of square frames, each containing N characters. In addition, the rows and columns were re-arranged, so that each was displayed in a separate box; Figure 7 shows the arrangement. Therefore, each character existed in two sets: one corresponding to the row, and the other corresponding to the column. Each set appeared and flashed on the screen at a fixed point in the center to help the subject to avoid eye movement (an eye tracker was used to track gaze positioning). The identification of the target character was based on the classification of the two sets in which the target character appeared. However, the system did not show great performance to compete with typical BCI spellers. A similar approach was already discussed before in [79]; however, in this study, two different arrangements for the sets were tested (without an eye tracker).

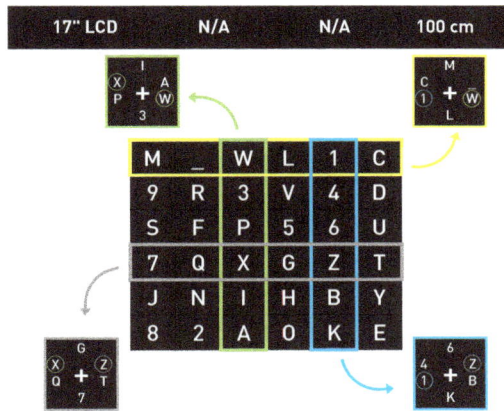

Figure 7. The GeoSpell as discussed in Aloise et al. [78], showing the group organization concept. The figure is modified from the original source [78].

This interface was declared to be a gaze-independent transformation of the matrix speller. Even though the accuracy was similar to the RCP, the typing speed was low compared to other spellers. This was mainly due to the number of different frames which were displayed in each trial.

The same team who presented the original Geospell system carried on further research on the same speller a couple of years later to investigate the unexpected low performance obtained in 2012 [103]. They compared the performance of the Geospell during covert attention with that of the Matrix Speller with overt attention. The aim of the study was to find an explanation, i.e., why the performance during covert attention was lower than during overt attention. The authors concluded that the overt attention modality was more accurate than the covert attention one, as it was cognitively more demanding. However, they also mentioned that, with some compensations, the Geospell could be equally or more accurate than the Matrix speller.

Modifications of the GeoSpell were developed and discussed in [82]: Motion-Covert GeoSpell (MCGS) and Covert GeoSpell (CGS). The purpose of this study was to investigate the performance of mVEPs for multi-objective gaze-independent BCIs. MCGS used motion-flash stimuli where characters appeared and moved a fixed distance to the edge of the screen during the presentation, while CGS only used the usual flash stimuli. Both systems were tested under the covert attention condition. The offline results showed that a higher P300 was evoked by the CGS compared to MCGS. This study concluded that mVEP could not enhance the performance of multi-objective gaze-independent BCIs regarding ERP.

4.2.5. Gaze-Independent Block Speller (GIBS)

The paradigm in [80] targeted the problems of covert attention and it was based on P300 BCI. The presented GUI had 30 characters, as shown in Figure 8. The symbols were grouped into four blocks which were located at the corners of the screen. Group 1 = [A B C D E F G], Group 2 = [H I J K L M N], Group 3 = [O P Q R S T U], Group 4 = [V W X Y Z 0 1]. The stimulation consisted in the flashing of the different blocks. The selected block was then expanded in the middle of the screen in a diamond shape (Figure 8). The second stage was presented by the flickering of the individual characters in the shape of a diamond. When the symbols were expanded to the center, they were larger and far apart to avoid crowding. The results showed that GIBS can be used without ocular movement. Moreover, by using bit rate analysis, the authors showed that GIBS could produce similar information transfer rates when compared to the standard Row-Column (RC) speller.

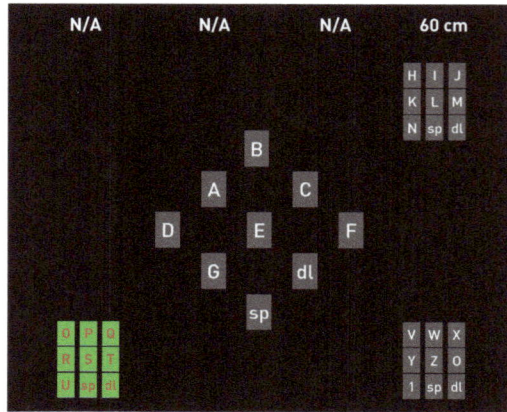

Figure 8. GIBS as discussed in [80]. Figure modified from [80].

GIBS was another gaze-independent P300-based speller, which also tried to avoid the matrix layout that has been proved to face some challenges. Being a two-phase system, fewer targets flickered per display.

4.2.6. Lateral Single Character Speller (LSC)

Pires et al. [81] proposed a lateral single character speller that was compared to other RC spellers; the layout reduced the effect of the local and remote distractors. Furthermore, the paradigm was expected to be more visually attractive and comfortable. The proposed Lateral Single Character Speller (LSC) speller, shown in Figure 9, contained the 26 letters of the alphabet and the 'spc' and 'del' commands. The 28 symbols flashed alternately and pseudo-randomly between the left and right fields in a lateral and symmetrical arrangement. The user had to focus only on one side of the screen, looking at one half of the display at a time. The target word for the copy task and the selected letters were shown in the middle of the arrangement, which required a short eye movement by the user.

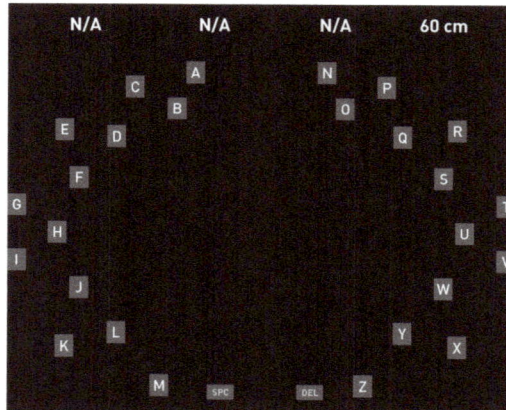

Figure 9. Lateral Single Character Speller, similar to [81]. Figure modified from [81].

As the interface divided the targets into two separated groups, it avoided the crowdedness problem as well as the adjacency-disturbance error, which occurred in the RCP because of the flashing of many letters, which were located close to each other. The paradigm showed better performance than the standard RC speller and it was effective with ALS patients and other patients with neuromuscular diseases (ITR 26.11 bits/min and accuracy 89.90%). From the questionnaires, the test subjects reported a preference towards the LSC, stating that it was more comfortable.

4.2.7. Hex-O-Spell as ERP

The Hex-O-Spell, a gaze-independent BCI speller that relies on imaginary movement, was first developed in 2006 by Blankertz et al. [39] and also presented in [104] (described in details in Section 4.4.1). This type of BCI has inspired many researchers to develop new BCI spellers. Here, we discuss some variations of the original Hex-O-Spell [39]. These variations were mainly developed to study the possibility of gaze-independent BCI speller systems which can be useful for late-stage ALS patients.

The first variation of the Hex-O-Spell was mentioned in [105], to be used as an ERP P300 BCI system, to test if there was a difference between the system's performance during covert attention and overt attention. In this study, a Hex-O-Spell, with minor changes with respect to its GUI, was compared to an adapted Matrix Speller. The modified Hex-O-Spell had circles around a central invisible hexagon instead of hexagons around a circle (similar to Figure 10a). The intensification was done with size changes. The size of the characters in the circle and the circle itself increased in turn, one by one. The Hex-O-Spell showed higher accuracy and a higher ERP response than the Matrix Speller in both covert and overt attention conditions.

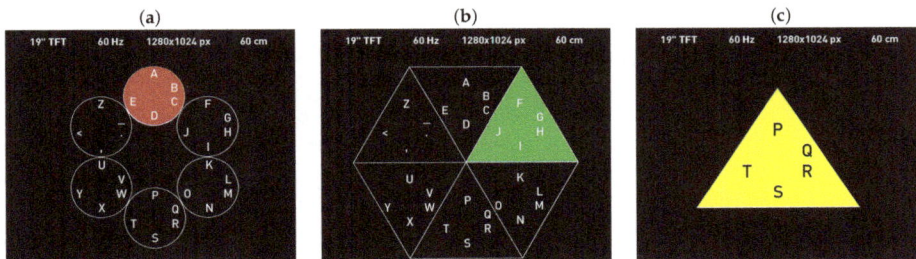

Figure 10. Three variations of the Hex-O-Spell with ERP for gaze-independent BCI studies. (a) Hex-O-Spell; (b) Cake Speller; (c) Center Speller. Figure modified from Treder et al., 2011 [83].

Other variations of the Hex-O-Spell utilizing ERP systems are the Cake Speller and Center Speller [83]. These two GUIs were developed to be compared with the Hex-O-Spell ERP in [105] for gaze-independent BCI spellers. During this study, a different intensification technique was used for the Hex-O-Spell ERP (Figure 10a). Instead of only changing the character size for intensification, the fill-in color of the circles was also changed, from the standard black background to one of six different colors (each circle had its own color). The Cake Speller was composed of a central hexagon, which was divided into six equal triangles, as shown in Figure 10b. Each triangle contained five characters, and the triangles were intensified one by one, by changing the fill-in color, using the different colors assigned to each triangle. Once one of the triangles was selected, the characters expanded and were distributed to the six triangles for the second stage of the selection.

The Center Speller is shown in Figure 10c. In this BCI speller, each group of characters (also five per group) was represented alone in the center of the screen, within a colored geometric shape. The combinations of letters and symbols within the geometric shape were called "elements". Each element was presented individually in the middle of the monitor with a unique color and inside a unique geometric shape. After the first stage of selection, the individual letters filled into the groups, in each element for the second-stage selection. To clarify, for each selection stage, six different displays were presented. The character intensification, in this case, occurred simply by the appearance of the element in the screen. A later study combined the Center Speller with the Hex-O-Spell, to create a test- and error-detection approach [84]. This paradigm can also be considered as an RSVP, which is discussed in the next section.

In [105], the Hex-O-Spell was transformed into an ERP system, to test whether ERP spellers could also be gaze-independent or not, with the aim to check if BCI-spellers could substitute eye tracker speller systems. Although the accuracy with covert attention was relatively low, it still proved that the ERP speller could operate with covert attention only and that accuracy could be improved in further studies. However, the Hex-O-Spell showed a higher performance during covert attention than the Matrix Speller. This concluded that the change in the design could improve the performance of the speller and grant a more effective control without the need for spatial attention. However, it still has to be tested on ALS patients. Other variations mentioned in [83] were also developed with the primary purpose of achieving gaze-independent spellers, such as the flashing in different colors as an intensification technique to elicit a stronger ERP signal. Although the typing rate was relatively slow, all the suggested interfaces showed high accuracy compared to other spellers (especially gaze-independent spellers). This could be treated as a proof of concept that ERP spellers can be effective without the need for gaze attention.

4.2.8. Rapid Serial Visual Presentation (RSVP)

RSVP was developed with the aim to form an efficient gaze-independent ERP speller. The paradigm was quite simple: individual characters appeared in the center of the screen in a randomized manner. The target letter evoked an ERP signal when it appeared. It was first presented in 2010 by Acqualagna et al. [85]. In this study, two variations were presented, a monochrome one and a colored one. A total of 30 characters (letters and punctuations) were presented. In the colored version, the characters were divided into three different colored groups: red [A B C D F G H I J −], green [K M N W E Q R S T +], blue [U V O X Y Z L P ! /]. To prevent symbols from clustering together frequently, pseudo-randomization was applied on the order of presentation, and, to allow for significant behavioral data to be obtained, the number of occurrences of the target symbol varied before and after each trial. The user had to look at the screen and count the number of times the target letter appeared in the middle of the display. The subjects showed better performance with the colored letters than with the monochrome ones. The accuracy of the RSVP speller outperformed both that of the Matrix speller and that of the Hex-O-Spell.

In 2011, Acqualagna and Blankertz [86] investigated three variants of the RSVP paradigm GUI, with different colors and different speeds of character representation. The performance of this paradigm was also tested online in 2013 [87].

Although the accuracy of the RSVP was very promising in all mentioned studies (around 94% on average), the ITR was lower than expected. The reason was that the user had to wait for the target letter to show up in between the rest of the 30 characters. This waiting wasted time and might have caused the user to feel bored or to lose focus. However, the system worked as a gaze-independent speller.

In [88], a new visual ERP-speller using N100 in addition to P300 was proposed. N100 is a type of visual evoked potential (VEP) which is induced by paying attention to the visual stimulus, and is not related to the oddball paradigm, making it difficult to use N100 alone for BCI. The authors in [88] claimed that this was the first time where N100 was used for BCI commands classification. In the proposed system, P300 and N100 were used separately and independently to determine the target character and to overcome the familiar challenges of an ERP speller. The GUI was similar to the standard 6 × 6 P300 speller with 36 commands: 26 letters (A–Z) and 10 numbers (0–9). However, the stimulus presentation was based on rapid visual presentation (RVP) to enable the implementation of the N100 into the BCI. Two BCI systems were developed, 2 × 2 and 2 × 3 matrices, which were presented as an RSVP stimulus. Each layout contained a group of the characters arranged in fixed positions. The assumption was that the user knew beforehand the position of the target letter. In the case of the 2×2 matrix, one position was left blank for three of the 12 different stimulation images in order to elicit the N100 signal. For the 2×3 matrix, nine stimulus images were used with similar blank positions. The input speed was faster than that of the P300 speller, as only nine stimulation sequences were required. The classification occurred by combining the P300 signal with the corresponding N100 from the blank position.

Eleven healthy 22–24-year-old males participated in this experiment, and performance comparisons between the two presented layouts and the P300 speller were carried out. For the 2×2, the average accuracy for the P300 was 63.1%, while the proposed speller in [88] showed an average accuracy of 74.7%. The average ITRs were 0.53 bpm and 0.70 bpm for the P300 and the proposed system, respectively. As for the 2×3 layout, the accuracies were 67.8% and 70.3%, and the ITRs were 0.60 bpm and 0.85 bpm for the P300 and the proposed system, respectively. The introduction of the N100 provided a one-stage selection and, as a result, it reduced user's fatigue and improved the accuracy of the system. However, it required the user to memorize the position of each letter beforehand, which might be difficult for some potential users.

4.3. SSVEP Spellers

An advantage of the SSVEP approach is that it does not require calibration or subject training. In addition, SSVEP spellers should be generally faster than P300 spellers, as no specific number of trials are required for them. A target can be selected as long as the signal is strong and stable enough to be detected by the software.

4.3.1. Bremen Speller

One of the earliest high-speed SSVEP-based BCI spellers is the Bremen-BCI speller [106]; a similar GUI is shown in Figure 11. In this study, a virtual diamond-shaped keyboard containing 32 characters was presented. The five boxes (four with arrows and one with the command "Select") were used to control the movement of a cursor which could move along the characters and select the desired target. Each of these boxes flickered with a certain frequency to elicit an SSVEP response. The letters were arranged according to their usage frequency in the English language. At the beginning of each trial and after each selection, the cursor was located by default in the middle, over the letter "E". The system gave audio feedback to the user, i.e., the system announced the selected letter out loud, so that the user or anyone nearby could hear it as a selection confirmation. Later in [28], a built-in dictionary was added to predict the desired words, as well as another type of feedback to notify the user about the selection, consisting in the size of the white boxes varying according to the power of the SSVEP signal, e.g., when the SSVEP signal increased, the size of the box increased, to notify the user that a selection was about to be made. Figure 12 shows the addition of the prediction module. It consisted of two different stages (layouts). The first stage was similar to the previous Bremen-BCI Speller with an extra sixth box with the command "Go" (Figure 12a). After the selection of at least two letters, a drop-down list of six words suggested from a dictionary appeared next to the "Go" command. If one of these choices was the desired word, the user had an option to select "Go". This action would lead to the next layout where each of the suggested words was presented in a flickering box (Figure 12b) and could be selected by the user to be written.

The Bremen-BCI Speller had gathered over the years a remarkable number of subjects. In addition to hundreds of tested healthy subjects, 37 participants were recruited during the RehaCare rehabilitation fair, eight of them with different disabilities [106]. Each participant took part in five different spelling tasks. The average ITR reported was 25.67 bits/min, with an accuracy of 93.27%, which indicated a competitive performance, especially for patients with neural malfunctions. Of note is that the experiments were carried out during a rehabilitation fair with a high level of noise and surrounding distractions. As for the Bremen Speller with the built-in dictionary, it showed a faster performance when compared to the original speller, with 32.71 bits/min and 29.98 bits/min, respectively. As another advantage, the dictionary implemented kept track of the most commonly used words. This feature speeded up the spelling by proposing the most often used words first. Also, it is worth mentioning that this speller was the first SSVEP-based speller with the option to predict words. After further improvements in signal processing, an average ITR of 61.70 bits/min, with a peak of 109.02 bits/min, was achieved with the Bremen-BCI speller, in a test with seven participants [107].

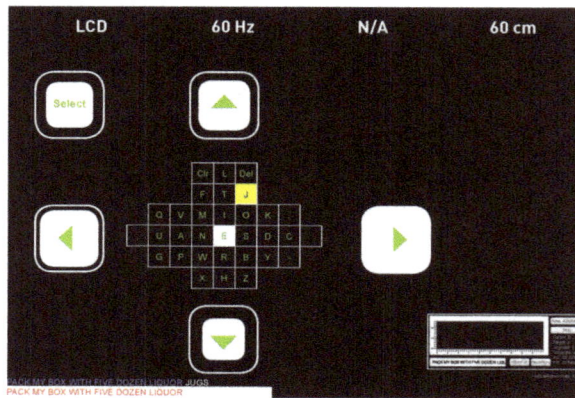

Figure 11. A similar GUI to the Bremen-BCI Speller during the selection of the right arrow, as the box size is increasing during selection [106]. Figure modified from [106].

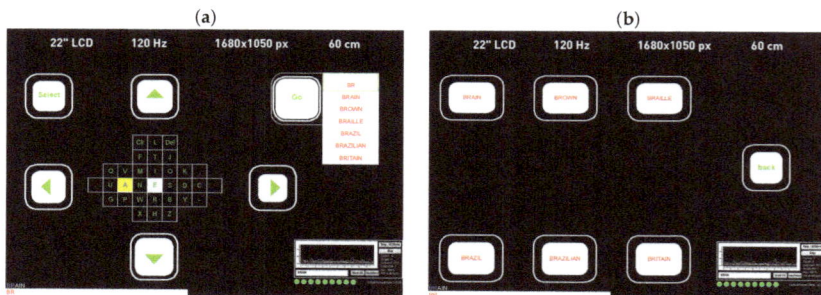

Figure 12. (**a**) The modification of the original Bremen-BCI speller when a build-in dictionary was added to it [28]; (**b**) The second stage of the GUI, where suggested words were presented to the user to choose the desired word.

4.3.2. Multi-Phase SSVEP Spellers

We presented a three-phase SSVEP speller in [29]. This study aimed to investigate the performance's differences of SSVEP-speller according to the subjects' age. The GUI consisted of four flashing white boxes with green characters or commands inside them, on a black background. Only the white boxes flashed while the green text was fixed. One of the boxes showed the command "delete". The other three boxes contained the letters of the alphabet. On the first screen, nine characters per box were displayed. When one box was selected, its content was spread over three boxes to form three characters per box. During the last selection phase, when a box was selected, the three boxes contained one letter each. In the second and the third stage, the "delete" box changed to "back" to give the option to go back to a previous layer in case of error. Every selection gave an audio feedback, naming the selected box. A similar approach was presented earlier in [30].

In [29], all subjects from two groups, a young age group and an older group, achieved control over the BCI system. The mean values of the young group were 98.49% accuracy with ITR of 27.36 bits/min, while the older group's accuracy was 91.13% with ITR of 16.10 bits/min. Although there was a significant difference between the performances of these two groups, the system was reliable with relatively high performances.

In another study, Cao et al. [31] proposed an SSVEP-based speller system with two phases. This speller would allow the input of 42 characters comprising of letters, digits, and symbols. Its user interface had three pages and 16 targets on each. Page turning was done via two boxes (buttons) which previewed the characters on hold on their corresponding page, aiding the user. Another two-phase SSVEP speller was discussed in [98] and compared to the Bremen speller. In this study, we presented a GUI with five boxes; each contained six alphabet characters and special symbols. Two other boxes were present, one containing the commands "delete" during the first window and "back" in the second window, and the other with the command "Clear", where the user could delete the whole word. When a box was selected, the content of this box was spread out to form one letter or symbol per box, in the second window. Another recent multi-stage SSVEP speller was presented in [32].

The two-phase SSVEP spellers may have a higher performance than the three-phase spellers, as fewer steps were needed for letter selection. In [98], comparing the two-stage SSVEP speller with the Bremen-BCI Speller, most subjects stated that the two-phase speller was more user-friendly than the Bremen Speller. However, the mean values for both spellers regarding ITR, accuracy, and time did not show any substantial difference for any of the tasks.

All the rest of the SSVEP mentioned studies proved remarkable performances, with the highest mean accuracy of 98.78% presented in [31] and mean ITR of 61.64 bits/min. However, none of the mentioned studies, except the one about the Bremen Speller, included MND patients as subjects for testing the system.

4.3.3. Multi-Target One-Phase SSVEP Spellers

Multi-Phase SSVEP spelling systems typically utilize a low number of distinct stimuli. The number of stimuli is anti-correlated to the number of phases. A low number of stimuli results in a low spelling speed, as classification and gaze-shifting phases of each phase are accumulated. Several groups, therefore, developed spelling applications that employ multiple stimuli simultaneously. This allows letter selection in a single step, resulting in much higher spelling speeds.

Wang et al. proposed a method to realize multiple SSVEP stimuli on computer screens [33]. The method was initially tested online with a virtual keypad consisting of 16 SSVEP target stimuli. The three subjects achieved an ITR of 75.4 bits/min, with an average accuracy of 97.2%.

Meanwhile, the methods were further improved and led to the highest ITR values reported for BCI spellers. As a result of refined classification methods and user-specific calibration data, Chen et al. achieved average ITRs of 267 bits/min and accuracy of 89.76%, employing 40 SSVEP targets [34]. The stimuli were arranged as a 5×8 matrix containing characters, numbers, and additional symbols.

Recently, Nakanishi et al. reported an average ITR of 325.33 bits/min in a cue-guided task using a 40-class speller, with an accuracy of 89.83% [89]. It was also stated that free spelling resulted in a slightly lower ITR (198.67 bits/min) and that inexperienced users required longer gaze shifts. In general, a higher number of targets in SSVEP-based BCI increases the spelling speed but also increases eye fatigue and target misclassification.

Multi-target BCI spellers have also been realized using the c-VEP paradigm. Spüler et al. achieved an average ITR of 143.95 bits/min with nine subjects [35]. The 32 target stimuli were arranged as a 4×8 matrix and were used to select letters, numbers, and underscore. Wei et al. tested a 48-target c-VEP system with four participants and achieved an ITR of 129 bits/min [36]. In c-VEP-based BCIs, all stimuli share the same circular shifted code pattern. Thus, the spelling accuracy requires precise timing between stimuli presentation and data acquisition.

Recently, Nagel et al. investigated the effect of monitor raster latency [108]. By correcting the raster latency, the distance between the most probable and the second most probable target was increased by 18.23%, resulting in a more reliable system.

4.3.4. RC SSVEP Speller

A dynamically optimized SSVEP brain–computer interface speller was presented in [90], which emulated the stimulation technique of a P300 Matrix Speller. A row/column (RC) paradigm was introduced into the SSVEP BCI to create an SSVEP speller with 36 items, flickering with only six frequencies, one for each element in a row or a column. A similar stimulation approach, which combines the P300 with SSVEP, was previously discussed in [91,92] as well. In [93], a SSVEP and P300 combination was also used; however, the P300 stimulus was different. The authors applied changes in the color, size, and rotation of the characters for stimulation.

4.3.5. FlashType^{TM}

The FlashType^{TM} [37] is one of the newest c-VEP BCI spellers. This type of speller does not rely on selecting individual letters like the previously discussed spellers. Instead, it controls the movement of a cursor and selected letters or symbols from a static keyboard. In the center of the displayed window, the keyboard had 28 visual targets; a row above showed the suggested character to be selected, and another row at the top showed the suggested words to be typed. The arrangement of the user interface was designed to utilize the majority of the screen and also to maximize the inter-stimuli distance. In the four corners of the screen, the stimuli were presented in four green/red 5 × 5 checkerboards with two patterns each. The shifting between the two patterns was according to a special pseudorandom binary code, which resulted in an induced Coded Visual Evoked Potential (c-VEP) signal. The four stimuli presented four different controls to the cursor: select, horizontal movement, vertical movement, and reverse. The horizontal and vertical controls pointed the movement of the cursor in the desired direction, while the select stimulus selected the target letter. The reverse stimulus moved the cursor in the opposite direction relative to the default direction in the horizontal or the vertical mode. The first character from the left on the Character Suggestions row was the cursor's default starting point. Then, the subject proceeded by moving the cursor vertically to select the looked-for row. The active row was marked in a yellow frame. After row selection, horizontal movement was required to select the wanted column; an active column was marked in a purple frame. The selection of the column resulted in the selection of the target letter, which fell in the intersection of the selected row and column.

Another mode of operation, which is still under further study, is the auto-scroll mode. This mode was developed to give minimum to no gaze-dependency. During auto scroll, only one stimulus was active, the select stimulus. The cursor moved automatically, stopping at each row and column in a specific order. All that the subject needed to do was to "select" the target while the cursor was pointing at it. Although that mode might be helpful for patients with no eye movement control, it is extremely slow.

This study reported notable advantages. The experiment was conducted using only one electrode to read the signal for the four stimuli which aided the system to be more user-friendly, as less preparation was required. It also classified the system as being relatively more portable compared to other BCIs which require eight or more electrodes. The performance results showed high accuracy and relatively fast typing for all three subjects. Moreover, a significant advantage of the static keyboard was that the characters could be substituted for characters of any language or even replaced by communication symbols. The predictive words option made the speller faster and more user-friendly. In addition, the speller did not require a lot of eye movements as the subject needed to only move attention to the four stimuli controlling the cursor and not to each character. Plus, the added mode of auto-scroll could be relatively slow, however, it would be beneficial for patients without oculomotor control. On the other hand, the interface was only tested by three healthy subjects, which is a small number compared to other studies, especially modern studies. In addition, none of the participants were NMD patients.

4.3.6. DTU BCI Speller (Technical University of Denmark)

DTU BCI is an SSVEP-based BCI developed by Vilic et al. [38]. The typing area divided the screen. On the left-hand side, there were seven stimuli. Each of them corresponded to a group of symbols, which was active during the first stage of spelling. During the second stage, the stimuli boxes on the right-hand side started flickering. These boxes presented the suggested words from the built-in language model dictionary. Another flickering box was situated at the bottom of the typing area. This box was always active. It permitted the user to voluntarily choose to switch from the first spelling stage to the next. Once a target was selected, it turned green for few seconds to give a feedback to the user. When a word was chosen, space was added, and the stimulation was active on the left-hand side again, starting automatically.

The users gave a positive feedback regarding the friendliness of the interface. As only three electrodes were used, the setup time was minimal, and the portability of the system was realistic. The added built-in dictionary for word prediction supported the users to reach faster typing speeds using this BCI speller. The average overall performance of this system was reasonable compared to other SSVEP spellers (ITR 21.94 bits/min and accuracy 90.81%).

4.4. MI Interfaces

In this section, we present different spellers which are based on MI. A unique feature of MI-based systems is that they are not dependent on any kind of external stimuli.

4.4.1. Hex-O-Spell

The Hex-O-Spell, a gaze-independent BCI speller that relies on imaginary movement, was first developed in 2006 by Blankertz et al. [39], also presented in [104]. It was inspired by a mobile device, which relies on the change of the device orientation for typing. The aim was to develop a working, synchronous BCI system, with the least number of controls possible (two) for 30 targets (26 letters + punctuations). The two controls were based on two mental states: imagined right-hand movement and imagined foot movement. As shown in Figure 13a, six hexagons were arranged eccentrically around a circle containing an arrow pointing out from the center towards the hexagons. The 30 characters were divided equally among the hexagons, five characters each. By imagining the right-hand movement or foot movement, the subject could rotate the arrow or select the hexagon that the arrow points to, which contains the target letter, respectively. In the second stage of selection (Figure 13b), the characters from the selected group spread out in a way that each letter or symbol occupied one of the hexagons. If an error was made during the first selection, the sixth (empty) hexagon gave the user the option to return to the first stage. Then, this two-step process was repeated to spell a complete word.

This was another gaze-independent speller, which might be appropriate for advanced-stage ALS patients. As this was an MI-based speller, which means the user has the ultimate control over the system (no external stimulation was necessary), the user had to practice using the system. As the user had more control over the system, fewer errors were expected to occur. The system's spelling speed was also dependent to some extent on the speed at which the user was controlling the speller. All in all, this was a state-of-the-art speller system with the advantages of using MI: no necessary stimulations and gaze-independency. However, the typical disadvantages of an MI system were also present, i.e., the extended training periods, the incurred fatigue, and the increased complexity of the data analysis.

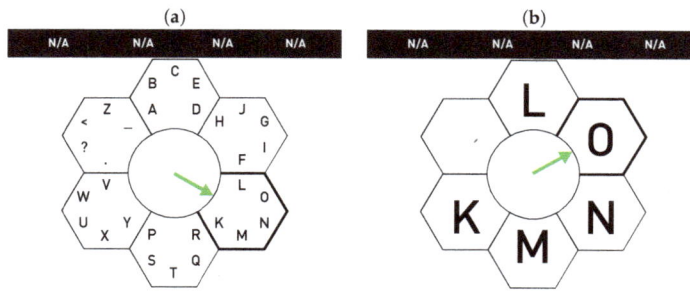

Figure 13. The GUI similar to Berlin Hex-O-Spell GUI, as shown and discussed in Blankertz et al., 2006 [39] (**a**) during the first stage of selection and (**b**) the second stage for selecting an individual letter. Figures modified from [104].

4.4.2. Oct-O-Spell

Another recent MI-based speller, the Oct-O-Spell was introduced in [40]. The GUI is similar to the Hex-O-Spell. The system was controlled asynchronously by a brain switch, which means that the user can turn the speller on and off by a specific brain signal. In the first phase, the interface showed an octagon divided equally into eight sections. These sections contained in total 26 letters, six letters, digits, or symbols, each. The second phase was dependent on the first stage selection. The sector selected during the first stage was unfolded across the eight sectors (there were only six characters in each section to unfold, the commands "Back" and "Delt" were added to have eight sectors in the second stage). For only two selections from the second stage, a third stage was implemented, to verify whether the user really wants to quit ("Yes" or "No"), or, after selecting the command "F1", to choose which symbol ("*", "@", "?", "+", "!", "#") should be written. Words suggestions appeared to the user after selecting several characters, which were simply selected by entering the corresponding number.

This interface showed a similar performance to other BCI spellers, especially hybrid BCI spellers. The interface was also tested without the predictive text. Although this case showed higher performance than the mode with the suggested words option, there was no significant difference between the two modes.

4.4.3. Other MI Spellers

D'albis et al. [94] presented a novel MI-based BCI speller. The GUI consisted of four rectangles at the edges of the screen. Three boxes (upper, right, and left) contained English characters. The fourth bottom box contained commands to help the user to control the speller, such as undo, delete, switch from letters to numbers, and quit the interface. Specific imaginary movements could be used to select each of the boxes. The left and right boxes could be selected by imagining the movement of the left and right arm, respectively. The upper box was activated by the movement of both arms, and the lower box by both legs. A predictive method was applied by them to lower the number of steps necessary for selection. This was done by enabling/disabling each character according to their probability to follow the already written text. When one of the boxes was selected, the enabled letters inside were extended for a single-character selection step.

Only three healthy subjects tested this paradigm and showed a humble performance; however, it proved the concept of the suggested interface. An additional advantage of this system was the embedded prediction module, which displayed suggested words for the user to select.

The GUI presented in [41] (also previously mentioned in [106] as Bremen speller) is a diamond-shaped interface divided into steps. The letters are arranged according to their usage frequency and into two layers. The layer 2 contained mostly numbers, three letters with the least usage frequency, and a delete symbol. By four different imaginary movements, the user could control a cursor and change between layers. The commands "up", "down", "left", "right", and "enter" were shown on the screen to move the cursor and select the target character.

The point of using SMR, instead of visual evoked potentials, was to avoid the uncomfortable stimulation technique. The system reported an average accuracy of 85% and proved the concept of operating the speller by using MI instead of visual stimuli. This latter point can attract more late-stages ALS patients to adopt this speller, as no eye movement was required.

4.5. Hybrid

To combine the advantages of two different systems, hybrid systems were developed to include more than one type of BCI or Human Machine Interface (HMI) paradigm, in general [109].

4.5.1. SSVEP + P300

In [95], another approach was presented, which can be considered as a T9 speller, given that the 36 letters and numbers were divided into nine groups, requiring only nine stimulation frequencies. It was a hybrid BCI system based on SSVEP and P300, with four different characters per group, which were flickering periodically in a random order. The flickering stimulus elicited an SSVEP while the "oddball paradigm" of random characters appearing in each group was responsible for the P300. Each character appeared in different color and was placed to improve the P300 signal and the performance of the system. The hybrid system showed superior performance when compared to either the SSVEP or theP300. The individual SSVEP and P300 only systems resulted in 89% and 90% accuracies, respectively, while the accuracy of the hybrid system was 93%. As for the ITR, it was 13.0, 19.9, and 31.8 bits/min for the SSVEP, P300, and the hybrid, respectively.

4.5.2. 60 Target Hybrid SSVEP/EMG (Electromyogram)

Lin et al. [96] created a 6×10 speller matrix on an LCD monitor, containing 60 stimuli. The 60 characters were divided into four equal groups, resulting in 15 characters per group flickering at 15 different frequencies. To select one of the groups, the user had to make a fist movement. For each group, a specific number of fist movements were required. Group 1, 2, 3, and 4 required zero, one, two, and three movements, respectively. After selecting the desired group, the users needed to select the target letter by gazing at it to elicit an SSVEP response.

This combination of SSVEP and EMG resulted in one of the highest mean ITR values of all the systems mentioned in this review, i.e., 90.9 bits/min, with a reasonable average accuracy of 85.80%. As neither SSVEP nor EMG requires training, the system was relatively easy to use for a wide group of users. However, as EMG requires actual physical movement (in this case, wrist movement), it limits the number of users who could profit from such a high-performing speller.

4.5.3. Consonant/Vowel Lists

In [97], a hybrid BCI speller system based on Motor Evoked Potential (MEP) (a type of MI) and P300 was presented. The idea was to use the MEP when there was a low number of targets and the P300 when there were more of them. The speller consisted of two lists, one containing the consonants and the other the vowels. The letters were arranged according to their probability of usage. By imaginary left- and right-hand movements, the user chose between the lists. Once a list was selected, the letters started flickering, and the user focused on the target letter for selection. After three spelled letters, suggested words appeared on the screen and could be selected by the user. The word printed on the screen was followed by a space. The system showed faster performance but reduced accuracy.

The reported results showed average ITR and accuracy levels compared to other studies mentioned in this review. However, the idea of dividing the letters into consonant and vowel lists might make it easier for users to spell and can achieve higher typing speed after a number of trials. It was also proven that the hybrid system achieved better performance than the individual P300 or MI.

This interface might be helpful for users who have difficulties focusing on the flashing Matrix speller. However, this speller was only tested by two healthy participants.

4.5.4. MI + P300

In [42], a self-paced hybrid BCI speller was introduced. It was a matrix speller based on Farwell and Donchin's speller, but it was controlled by an MI switch. A matrix with the size of 6 × 7, including 26 English letters, 10 numbers, comma ",", dot ".", "Del", and "Exit", was initially turned OFF at the start of the experiment. The user had to intentionally change his/her state of mind by MI to turn the system ON and start the flashing of the stimuli. The hybrid speller showed exciting results, the average classification accuracy and the ITR were 92.93%, and 41.23 bits/min, respectively. These results are relatively within the range of the commonly tested P300 matrix spellers, but, for this experiment, the user could control when to start and when to stop spelling intentionally, without affecting the performance of the speller.

5. Discussion

In addition to the developments presented in this review paper, various modalities have been presented throughout the years with the primary aim to improve the quality of life of users with disabilities. Eye tracking devices are a successful example, which are commercially available off the shelf for users and also used for further development of new applications for MND patients [110]. Many studies have been carried out to compare and/or combine BCI systems with eye trackers. One of the most recent studies is our study [111], which merges both systems to develop a hybrid speller combining the advantages of both systems in a competent speller. More information about eye tracking and the comparison between eye tracker spellers and BCIs can be found in [111].

Other systems have a special chin joystick, which could assist the user to manipulate an assistive robot, as presented in [112]. To mention other control modes, in [113], computer and assistive devices were controlled by the tongue, and, in [114], infrared sensors were used to detect head movement to control a computer mouse. All these methodologies are very useful and can benefit many people with disabilities. Although some of these methods can be faster and more accurate than BCI, when applied to spelling applications, they restrict the number of potential users, as they are only beneficial for patients who still maintain some motor control. BCIs usually require minimal or no muscular movement, qualifying BCI systems to be suitable for a wider group of users.

Spelling applications are our key focus of this review. Nevertheless, it is worth mentioning that BCIs are applied for a lot of other purposes, e.g., robot control [115–117], wheelchair control [118,119], web browsing [120], general control of an operating system with a virtual keyboard, as presented in [121], gaming [122], every-day electronic device control [123–125], and ADHD attention training [126], just to mention a few. A lot of these applications are also useful for people with neuromuscular malfunctions, and a number of them can be suitable for healthy users.

Although BCIs have a lot of advantages and benefits, there is room for improvements. Usually, BCI systems require some time and help to be set up. It is tricky for a person with no BCI experience to set up a commonly used BCI system. One of the reasons is the setup of the electrodes. The used electrodes need to be fixed at specific positions, and it is also essential to apply electrolyte gel correctly for EEG-based measurement. As an example, in [127,128], other types of electrodes and setup methods were proposed. Another challenge is the portability. Researchers have been working lately to overcome this barrier to achieve a more portable BCI system; just to mention some examples: [128–131].

In addition to working on the above-stated challenges, researchers always aim for BCI systems with better performance. Here, we discuss the developments made towards this goal in the aspect of GUI changes. Many articles have been published with respect to other properties, especially data processing and analysis of a BCI. Much more papers were presented in this field (examples: [132–135]).

Back to the main topic of discussion, the BCI spellers, various interfaces and systems are reported in various literature libraries. As each presented system had its own variables, parameters, and conditions, it is not possible to carry out an objective comparison between different GUI spellers. However, we can still discuss some advantages and disadvantages of the systems described in the previous section of this review. Tables 2–4 summarize the studies mentioned above, while stating the main specifications of each system, such as accuracy, information transfer rate (ITR), and the number and type of subjects who participated in the studies. Table 2 includes all the variations and studies which were built on the original Matrix Speller. Table 3 presents a summary of the performance of other P300-based spellers which were not directly inspired by the Matrix Speller, and, finally, Table 4 presents the performance of the above-mentioned SSVEP-based spellers, MI spellers, and hybrid spellers. We can see from these tables that most of the developments made during the decade are based on P300 BCIs, especially on the Matrix speller GUI.

Table 2. Summary of all spellers discussed in this review which are based on the P300 Matrix Speller.

Topic/Speller Name	Reference		Subjects	Mean ITR/Typing Speed	Mean Accuracy
Matrix Speller	[7]	Farwell and Donchin 1988	4 healthy	12 bits/min	95.0%
Stimuli Variations	[43]	Yeom et al. 2014a	4 healthy	66.3 bits/min	64.7%
	[44]	Obeidat et al. 2015	14 healthy	13.7 bits/min	93.3%
	[45]	Liu et al. 2010	4 healthy	rotation stimuli: 35.8 bits/min	rotation stimuli: 89.06%
	[46]	Shi et al. 2012	7 healthy	SBP433: 26.8 bits/min	99.7%
	[47]	Eom et al. 2013	5 healthy	13.5 s/char	79.2%
	[48]	Jin et al. 2010	8 healthy	14.8 bits/min	92.9%
	[49]	Polprasert et al. 2013	10 healthy	23.82 bits/min	84.0%
Familiar Faces and symbols	[50]	Kaufmann et al. 2011	20 healthy	N/A	Max 100%
	[51]	Li et al. 2015a	17 healthy	N/A	N/A
	[52]	Li et al. 2015b	12 healthy	39.0 bits/min	86.1%
	[53]	Kaufmann and Kübler 2014	8 healthy	~80 bits/min	81.25%
	[54]	Yeom et al. 2014b	15 healthy	RASP-F: 53.7 bits/min RASP: 32.8 bits/min	84.0% 90.7%
	[56]	Kathner et al. 2015	18 healthy + 1 LIS	15.5–16.2 bits/min	94–96%
Variation of letters arrangement	[57]	Ahi et al. 2011	14 healthy	55.32 bits/min	87.14%
	[58]	Li et al. 2011	10 healthy + 10 NMD	N/A	79.7–28.7%
	[59]	Jin et al. 2012	9 healthy	18-P: 29.9 bits/min 21-P: 27.1 bits/min	18-P: 93.3% 21-P: 94.8%
	[60]	Sakai and Yagi 2011	9 healthy	N/A	N/A
Matrix Speller with Prediction	[62]	Ryan et al. 2011	24 healthy	17.71 bits/min	84.88%
	[63]	Kaufmann et al. 2012	20 healthy	Max 25 bits/min	>70%
	[64]	Akram et al. 2013	4 healthy	26.1 s/char	77.5%
	[65]	Akram et al. 2014	10 healthy	26.13 s/char	77.14%
Other languages	[66]	Minett et al. 2010	30 healthy	14.5 bits/min	> 60%
	[67]	Minett et al. 2012	24 healthy	4.23 bits/min	82.8%
	[68]	Yu et al. 2016	10 healthy	39.2 bits/min	92.6%
	[69]	Kabbara et al. 2015	11 healthy	N/A	88–95%
	[70]	Lee et al. 2011	3 healthy	N/A	100% after training
	[71]	Yamamoto et al. 2014	4 healthy	N/A	93%
	[72]	Ikegami et al. 2014	7 ALS patients + 7 healthy	N/A	ALS: 24%, 55% healthy: 55%, 83%
3D Blocks Matrix Speller	[73]	Noorzadeh et al. 2014	16 healthy	N/A	~90% with 5 repetitions

Table 3. Summary of all other P300-based spellers which are not directly related to the Matrix Speller.

Topic/Speller Name	Reference		Subjects	Mean ITR/Typing Speed	Mean Accuracy
Chroma Speller	[74]	Acqualagna et al. 2013	9 healthy	1.4 char/min	88.4%
T9	[76]	Ron-Angevin et al. 2015	11 healthy + 1 with ALS	N/A	N/A
	[75]	Akram et al. 2015	10 healthy	26.125 s/char	N/A
Checkerboard Paradigm	[77]	Postelnicu and Talaba 2013	10 healthy	21.74 bits/min	90.63%
Geospell	[79]	Liu et al. 2011	8 healthy	1.38 char/min	RP: 87.8% FP: 84.1%
	[78]	Aloise et al. 2012	10 healthy	1.86 char/min	78%
	[82]	Zhou et al. 2016	10 healthy	N/A	N/A
GIBS	[80]	Pires et al. 2011	4 healthy	16.67 bits/min	96.02%
LSC Speller	[81]	Pires et al. 2012	10 healthy + 7 ALS + 5CP + 1 DMD + 1 SCI	26.11 bits/min	89.9%
Hex-O-Spell with ERP	[83]	Treder et al. 2011	13 healthy	2 char/min	Hex-O-Spell: 90.4% Cake Speller: 88.0% Center Speller: 97.0%
	[84]	Schmidt et al. 2012	11 healthy	2.75 char/min	89.1%
Rapid serial visual presentation RSVP	[87]	Acqualagna and Blankertz 2013	12 healthy	1.43 char/min	94.8%
	[85]	Acqualagna et al. 2010	9 healthy	N/A	90%
	[86]	Acqualagna and Blankertz 2011	12 healthy	2 char/min	94.8%
	[88]	Sato and Washizawa 2016	11 healthy	2 × 2: 0.70 bits/s 2 × 3: 0.85 bits/s	2 × 2: 74.4% 2 × 3: 70.3%

Table 4. Summary of the spellers discussed in this review which are based on SSVEP, MI, and Hybrid system.

Topic/Speller Name	Reference		Subjects	Mean ITR/Typing Speed	Mean Accuracy
Bremen Speller	[28]	Volosyak et al. 2011	7 healthy	32.71 bits/min	Correct spelling only
Multi-Phase Spellers	[29]	Volosyak et al. 2017	20 healthy	group A: 27.36 bits/min group B: 16.10 bits/min	group A: 98.49% group B: 91.13%
	[30]	Cecotti 2010	8 healthy	37.62 bits/min	92.25%
	[31]	Cao et al. 2011	4 healthy	61.64 bits/min	98.78%
	[32]	Ansari and Singla 2016	20 healthy	13 chars/min	96.04%
Multi-Target One-Phase Spellers	[33]	Wang et al. 2010	3 healthy	75.4 bits/min	97.2%
	[34]	Chen et al. 2015	12 healthy	4.45 bits/min	91.04%
	[89]	Nakanishi et al. 2018	20 healthy	325.33 bits/min	89.83%
	[35]	Spüler et al. 2012	9 healthy	143.95 bits/min	96.18%
	[36]	Wei et al. 2017	4 healthy	129.58 bits/min	90.5%
	[90]	Yin et al. 2015b	11 healthy	41.08 bits/min	~95%
RC SSVEP Speller	[91]	Yin et al. 2013	12 healthy	56.44 bits/min	93.85%
	[92]	Yin et al. 2014	14 healthy	RC: 53.06 bits/min SL: 44.7 bits/min	N/A
	[93]	Yin et al. 2015a	13 healthy	50.41 bits/min	95.18%
Flash-Type Speller	[37]	Nezamfar et al. 2016	3 healthy	6.2–11 s/char	95.5–97%
DTU BCI Speller	[38]	Vilic et al. 2013	9 healthy	21.94 bits/min	90.81%
Hex-O-Spell	[39]	Blankertz et al. 2006	2 healthy	max 7.6 char/min	error free measurements
Oct-O-Spell	[40]	Cao et al. 2017	3 healthy	Non-PTE: 69.16 bits/min PTE: 62.39 bits/min	Non-PTE: 98.3% PTE: 96.6%
Other MI Speller	[94]	D'Albis et al. 2012	3 healthy	max 3 char/min	average N/A
	[41]	Jingwei et al. 2011	5 healthy	N/A	85.0%
SSVEP+P300	[95]	Chang et al. 2016	10 healthy	31.8 bits/min	93%
SSVEP+EMG	[96]	Lin et al. 2016	10 healthy	90.9 bits/min	85.8%
Consonant/Vowels list	[97]	Roula et al. 2012	2 healthy	11 s/char	70%
MI+P300	[42]	Yu et al. 2016	11 healthy	41.23 bits/min	92.93%

Another reason why the performance varies from one study to another is the use of different resources. Different teams utilize different hardware. There are several available bioamplifiers on the market as well as a variety of EEG caps and electrodes. Additionally, researchers usually build the software which is suitable for them. Some develop their own software and others would make use of available tools on the market. These parameters affect the performance greatly. A first step to conduct a subjective comparison between different BCI systems is to make sure that the same hardware and software are being implemented.

It is also noticeable from the tables that there are performance variations between different systems in the same category. A main reason behind this is that the ITR and accuracy are calculated differently for different interfaces, which means that it is almost impossible to carry out an objective evaluation or comparison between the systems. In some cases, it is obvious that the improvement in the spelling speed affected the accuracy negatively or vice versa. For example, the Center Speller or the RSVP spellers (see Table 2 for more details) were specifically designed to achieve high accuracy and gaze-independent spelling. They successfully achieved their goals; however, the typing speed was affected significantly.

In other cases, a typing mistake requires time for correction, affecting the spelling speed. To eliminate this negative effects, researchers integrated Error-related Potentials (ErrPs) into BCI spellers. The ErrP signal is generated 50–100 ms after an error is detected by the user. The error might be due to a human error from the user's side, or it can be that the machine behaved differently from what the user expected. In [84,136,137], the ErrP was used to automatically detect and delete the errors in a BCI speller. The merging of ErrP with P300 aimed to increase the accuracy of the speller, while avoiding affecting the spelling speed, as the correction was done automatically. In our opinion, accuracy is more important for the subjects. During many experiments, we observed that the subjects were much more frustrated by typing mistakes than by a slow-performing system.

P300 is very popular among BCI researchers because of its relatively high ITR and the minimal user training required (compared to MI). However, in general, P300 spellers have several disadvantages. As the number of commands increases in a P300 speller, the number of trials increases as well, leading to a slower performance. As the feature extraction mostly depends on identifying the point of intersection of which row and column elicited the signal, at least two flashes are required for each target, which, again, increases the classification times. Although scientists have been working lately to develop gaze-independent P300 spellers, most of these systems require visual attention or even gaze shifting. The gaze shift dependency might not be applicable for patients with severe paralysis. It is worth noting that the modifications developed and applied to the original Matrix Speller paradigm showed better performance than that of the traditional RCP, especially when tested with MND patients. Some researchers worked on improving the performance of P300 spellers by developing classification algorithms to build an asynchronous P300-based BCI. As an example, the team in [138] combined a P300 speller similar to the Geospell (already mentioned in Section 4.2.4), which was gaze-independent, with an asynchronous algorithm. However, these results were not significantly different. Later, in [139], the group achieved promising results by embedding a self-calibration module to the system. This improvement included an algorithm which automatically recalibrated the parameters of the classifier and adjusted them according to the personal performance of the user. Accordingly, the system could adjust the parameters to attain the optimum accuracy and typing speed.

Some of these challenges are overcome in other BCI paradigms. For example, SSVEP does not require a minimum number of flashes to elicit a response. Thus, it can be used to achieve faster spellers. It also requires no training at all on the user's side. On the other hand, it has been observed that some participants have a low SSVEP response, which is almost impossible to detect and use as a control signal when a high number of stimuli is presented. As for MI, once the user receives the required training, the system can achieve impressive results. However, the training might take a long time.

The recently developed mVEP paradigm overcomes several other challenges which were faced during the development of other BCI systems. It is elicited entirely by the motion and the motion behavior of the visual stimulus. Thus, such system is not sensitive to contrast, illumination, color, or size of the stimulus. An mVEP-based paradigm does not require any previous training as well.

Another approach to avoid the gaze dependency problem is to utilize other sensory modalities than vision. As mentioned in the introduction, auditory and tactile ERP researches have been published with promising results. In [140], the Farwell and Donchin P300 speller was modified into a visual–auditory speller, which showed similar performance compared to the visual paradigm. In [141], an auditory matrix speller was presented, where the authors used natural animal sounds as stimuli. The system was tested by impaired subjects and resulted in a relatively high performance. However, these systems still require a lot of training and familiarization.

An example of a tactile ERP system is the Braille-like system developed in [142] as a P300-based tactile BCI system. Two or four different types of tactile stimuli were assigned to intentions, such as "yes" and "no", "right" and "left", or "up" and "down." The stimulation was a tactile stimulator using a piezoelectric actuator, used with the braille system, consisting of eight cylindrical pins which could be pressed lightly against the fingertips of the user. The number and the positions of the pins moving up defined the stimulus. Another recent study [143] combined tactile and auditory stimuli to form a hybrid BCI system. Although the performance of BCI spellers based on tactile and/or auditory stimulus (a stimulus other than vision) is still not better than the visual stimuli-based BCI spellers, specifically when used with healthy participants, they are of great importance when considering patients with eyesight problems or in late stages of ALS.

It is evident from Tables 2–4, that most of the studies were conducted using the P300 paradigm. This can be due to its popularity, its many advantages, and the fact that it has been studied for many years. This can guide us to deduce that there are more development opportunities in the other BCI paradigms. Overall, researchers have been working for many years to develop efficient BCI communication applications, which are safe, affordable, reliable, easy to setup, easy to use, and achieving a fast communication speed. From what we discussed above, BCI spellers can be a suitable option for people who have no other means of communication with their surroundings. However, BCI spellers are not fast enough, compared to other regular communication methods, like typing or speaking, neither as fast as other systems, like spellers based on eye trackers. Also, BCIs are relatively complicated to set up and require specific skills. A typical BCI system is not considered to be very portable. Some of the spellers discussed above aimed to tackle these challenges and gaps, but still more developments are required. Combining different BCI systems or combining BCI with other non-BCI systems can also lead to promising results. As already discussed, some spellers merged P300 with SSVEP, and some others combined them with other systems, like eye trackers. The results achieved by the hybrid systems were usually faster and/or more accurate. The general aim is to achieve a BCI speller which is as fast as other communication methods, easy to carry around and set up, comfortable to use for short terms and on the long run, and suitable for the broadest range of users.

6. Conclusions

All the systems discussed above were studied and presented with the aim to improve BCI spellers. Throughout the years, scientists have worked on spelling systems to make them faster, more accurate, more user-friendly, and, most of all, able to compete with traditional communication methods. On the other hand, a lot of gaps are still to be closed to achieve efficient BCI spellers. More emphasis needs to be given to GUI design to satisfy the needs of the end-users. In addition, more testing with patients is required. From the summary tables, we can see that only five systems were tested by afflicted subjects. BCI spellers provide a practical and efficient way for people who cannot communicate through traditional methods to be able to participate in their social lives and careers. The different GUIs described and discussed during this review, as well as the other different systems mentioned, may provide an inspiring starting point for further studies and improvements.

Acknowledgments: This research was supported by the European Fund for Regional Development (EFRD—or EFRE in German) under Grants GE-1-1-047 and IT-1-2-001.

Brain Sci. **2018**, *8*, 57

Author Contributions: A.R. and I.V. conceived and designed the review; all authors wrote the paper.

Conflicts of Interest: The authors declare no conflict of interest.

References

1. Hanagasi, H.A.; Gurvit, I.H.; Ermutlu, N.; Kaptanoglu, G.; Karamursel, S.; Idrisoglu, H.A.; Emre, M.; Demiralp, T. Cognitive impairment in amyotrophic lateral sclerosis: Evidence from neuropsychological investigation and event-related potentials. *Brain Res. Cogn. Brain Res.* **2002**, *14*, 234–244. [CrossRef]
2. Majaranta, P.; Räihä, K.-J. Twenty years of eye typing. In Proceedings of the Symposium on Eye Tracking Research & Applications—ETRA'02, New Orleans, LA, USA, 25–27 March 2002; pp. 15–22.
3. Sharma, R.; Hicks, S.; Berna, C.M.; Kennard, C.; Talbot, K.; Turner, M.R. Oculomotor dysfunction in amyotrophic lateral sclerosis: A comprehensive review. *Arch. Neurol.* **2011**, *68*, 857–861. [CrossRef] [PubMed]
4. Wolpaw, J.R.; Birbaumer, N.; McFarland, D.J.; Pfurtscheller, G.; Vaughan, T.M. Brain-computer interfaces for communication and control. *Clin. Neurophysiol.* **2002**, *113*, 767–791. [CrossRef]
5. Cecotti, H. Spelling with non-invasive Brain-Computer Interfaces—Current and future trends. *J. Physiol. Paris* **2011**, *105*, 106–114. [CrossRef] [PubMed]
6. Wolpaw, J.; Wolpaw, E.W. *Brain-Computer Interfaces: Principles and Practice*; Oxford University Press: New York, NY, USA, 2012.
7. Farwell, L.A.; Donchin, E. Talking off the top of your head: Toward a mental prosthesis utilizing event-related brain potentials. *Electroencephalogr Clin. Neurophysiol.* **1988**, *70*, 510–523. [CrossRef]
8. Felgoise, S.H.; Zaccheo, V.; Duff, J.; Simmons, Z. Verbal communication impacts quality of life in patients with amyotrophic lateral sclerosis. *Amyotroph. Lateral Scler. Frontotemporal Degener.* **2016**, *17*, 179–183. [CrossRef] [PubMed]
9. Wolpaw, J.R.; Birbaumer, N.; Heetderks, W.J.; McFarland, D.J.; Peckham, P.H.; Schalk, G.; Donchin, E.; Quatrano, L.A.; Robinson, C.J.; Vaughan, T.M. Brain-computer interface technology: A review of the first international meeting. *IEEE Trans. Rehabilit. Eng.* **2000**, *8*, 164–173. [CrossRef]
10. Shannon, C.E. A mathematical theory of communication. *ACM SIGMOBILE Mob. Comput. Commun. Rev.* **2001**, *5*, 3–55. [CrossRef]
11. Rao, R.P.N. *Brain-Computer Interfacing: An Introduction*; Cambridge University Press: Cambridge, UK, 2013.
12. Sutton, S.; Braren, M.; Zubin, J.; John, E.R. Evoked-potential correlates of stimulus uncertainty. *Science* **1965**, *150*, 1187–1188. [CrossRef] [PubMed]
13. McCarthy, G.; Donchin, E. A metric for thought: A comparison of P300 latency and reaction time. *Science* **1981**, *211*, 77–80. [CrossRef] [PubMed]
14. Hill, N.J.; Lal, T.N.; Bierig, K.; Birbaumer, N.; Schölkopf, B. An auditory paradigm for brain-computer interfaces. In *Advances in Neural Information Processing Systems*; NIPS Foundation: Vancouver, BC, Canada, 2005; pp. 569–576.
15. Höhne, J.; Tangermann, M. Towards user-friendly spelling with an auditory brain-computer interface: The charstreamer paradigm. *PLoS ONE* **2014**, *9*, e98322. [CrossRef] [PubMed]
16. Brouwer, A.M.; van Erp, J.B. A tactile P300 brain-computer interface. *Front. Neurosci.* **2010**. [CrossRef] [PubMed]
17. Van der Waal, M.; Severens, M.; Geuze, J.; Desain, P. Introducing the tactile speller: An ERP-based brain-computer interface for communication. *J. Neural Eng.* **2012**, *9*, 045002. [CrossRef] [PubMed]
18. Bin, G.; Gao, X.; Wang, Y.; Hong, B.; Gao, S. VEP-based brain-computer interfaces: Time, frequency, and code modulations [Research Frontier]. *IEEE Comput. Intell. Mag.* **2009**, *4*, 22–26. [CrossRef]
19. Snyder, A.Z. Steady-state vibration evoked potentials: Description of technique and characterization of responses. *Electroencephalogr. Clin. Neurophysiol./Evoked Potentials Sect.* **1992**, *84*, 257–268. [CrossRef]
20. Guo, F.; Hong, B.; Gao, X.; Gao, S. A brain–computer interface using motion-onset visual evoked potential. *J. Neural Eng.* **2008**, *5*, 477. [CrossRef] [PubMed]
21. Heinrich, S.P. A primer on motion visual evoked potentials. *Documenta Ophthalmol.* **2007**, *114*, 83–105. [CrossRef] [PubMed]
22. Kuba, M.; Kubová, Z. Visual evoked potentials specific for motion onset. *Documenta Ophthalmol.* **1992**, *80*, 83–89. [CrossRef]

23. Probst, T.; Plendl, H.; Paulus, W.; Wist, E.; Scherg, M. Identification of the visual motion area (area V5) in the human brain by dipole source analysis. *Exp. Brain Res.* **1993**, *93*, 345–351. [CrossRef] [PubMed]
24. Skrandies, W.; Jedynak, A.; Kleiser, R. Scalp distribution components of brain activity evoked by visual motion stimuli. *Exp. Brain Res.* **1998**, *122*, 62–70. [CrossRef] [PubMed]
25. Jasper, H.; Penfield, W. Electrocorticograms in man: Effect of voluntary movement upon the electrical activity of the precentral gyrus. *Arc. Psychiatr. Nervenkrankh.* **1949**, *183*, 163–174. [CrossRef]
26. Moher, D.; Liberati, A.; Tetzlaff, J.; Altman, D.G.; Group, P. Preferred reporting items for systematic reviews and meta-analyses: The PRISMA statement. *PLoS Med.* **2009**, *6*, e1000097. [CrossRef] [PubMed]
27. Müller-Putz, G.R.; Scherer, R.; Pfurtscheller, G.; Rupp, R. Brain-computer interfaces for control of neuroprostheses: From synchronous to asynchronous mode of operation. *Biomed. Tech. (Berl.)* **2006**, *51*, 57–63. [CrossRef] [PubMed]
28. Volosyak, I.; Moor, A.; Graser, A. A dictionary-driven SSVEP speller with a modified graphical user interface. In Proceedings of the 11th International Conference on Artificial Neural Networks Conference on Advances in Computational Intelligence, Torremolinos-Málaga, Spain, 8–10 June 2011; Volume I, pp. 353–361.
29. Volosyak, I.; Gembler, F.; Stawicki, P. Age-related differences in SSVEP-based BCI performance. *Neurocomputing* **2017**, *250*, 57–64. [CrossRef]
30. Cecotti, H. A self-paced and calibration-less SSVEP-based brain-computer interface speller. *IEEE Trans. Neural Syst. Rehabil. Eng.* **2010**, *18*, 127–133. [CrossRef] [PubMed]
31. Cao, T.; Wang, X.; Wang, B.; Wong, C.M.; Wan, F.; Mak, P.U.; Mak, P.I.; Vai, M.I. A high rate online SSVEP based brain-computer interface speller. In Proceedings of the 2011 5th International IEEE/EMBS Conference on Neural Engineering, Cancun, Mexico, 27 April–1 May 2011; pp. 465–468.
32. Ansari, I.A.; Singla, R. BCI: An optimised speller using SSVEP. *Int. J. Biomed. Eng. Technol.* **2016**, *22*, 31–46. [CrossRef]
33. Wang, Y.; Wang, Y.T.; Jung, T.P. Visual stimulus design for high-rate SSVEP BCI. *Electron. Lett.* **2010**, *46*, 1057–1058. [CrossRef]
34. Chen, X.; Wang, Y.; Nakanishi, M.; Gao, X.; Jung, T.-P.; Gao, S. High-speed spelling with a noninvasive brain–computer interface. *Proc. Natl. Acad. Sci. USA* **2015**, *112*, E6058–E6067. [CrossRef] [PubMed]
35. Spüler, M.; Rosenstiel, W.; Bogdan, M. Online adaptation of a c-VEP brain-computer interface (BCI) based on error-related potentials and unsupervised learning. *PLoS ONE* **2012**, *7*, e51077. [CrossRef] [PubMed]
36. Wei, Q.; Gong, H.; Lu, Z. Grouping modulation with different codes for improving performance in cVEP-based brain–computer interfaces. *Electron. Lett.* **2017**, *53*, 214–216. [CrossRef]
37. Nezamfar, H.; Mohseni Salehi, S.S.; Moghadamfalahi, M.; Erdogmus, D. FlashType^{TM}: A Context-Aware c-VEP-Based BCI Typing Interface Using EEG Signals. *IEEE J. Sel. Top. Signal Process.* **2016**, *10*, 932–941. [CrossRef]
38. Vilic, A.; Kjaer, T.W.; Thomsen, C.E.; Puthusserypady, S.; Sorensen, H.B.D. DTU BCI speller: An SSVEP-based spelling system with dictionary support. In Proceedings of the 2013 35th Annual International Conference of the IEEE Engineering in Medicine and Biology Society (EMBC), Osaka, Japan, 3–7 July 2013; pp. 2212–2215.
39. Blankertz, B.; Dornhege, G.; Krauledat, M.; Schröder, M.; Williamson, J.; Murray-Smith, R.; Müller, K.-R. The Berlin Brain-Computer Interface presents the novel mental typewriter Hex-o-Spell. In Proceedings of the 3rd International Brain-Computer Interface Workshop and Training Course, Graz, Austria, 21–24 September 2006; pp. 108–109.
40. Cao, L.; Xia, B.; Maysam, O.; Li, J.; Xie, H.; Birbaumer, N. A Synchronous Motor Imagery Based Neural Physiological Paradigm for Brain Computer Interface Speller. *Front. Hum. Neurosci.* **2017**, *11*, 274. [CrossRef] [PubMed]
41. Jingwei, Y.; Jun, J.; Zongtan, Z.; Dewen, H. SMR-Speller: A novel Brain-Computer Interface spell paradigm. In Proceedings of the 2011 3rd International Conference on Computer Research and Development, Shanghai, China, 11–13 March 2011; pp. 187–190.
42. Yu, Y.; Jiang, J.; Zhou, Z.; Yin, E.; Liu, Y.; Wang, J.; Zhang, N.; Hu, D. A Self-Paced Brain-Computer Interface Speller by Combining Motor Imagery and P300 Potential. In Proceedings of the 2016 8th International Conference on Intelligent Human-Machine Systems and Cybernetics (IHMSC), Hangzhou, China, 27–28 August 2016; pp. 160–163.

43. Yeom, S.K.; Fazli, S.; Lee, S.W. P300 visual speller based on random set presentation. In Proceedings of the 2014 International Winter Workshop on Brain-Computer Interface (BCI), Jeongsun-kun, Korea, 17–19 February 2014; pp. 1–2.

44. Obeidat, Q.T.; Campbell, T.A.; Kong, J. Introducing the Edges Paradigm: A P300 Brain–Computer Interface for Spelling Written Words. *IEEE Trans. Hum.-Mach. Syst.* **2015**, *45*, 727–738. [CrossRef]

45. Liu, Y.; Zhou, Z.; Hu, D. Comparison of stimulus types in visual P300 speller of brain-computer interfaces. In Proceedings of the 2010 9th IEEE International Conference on Cognitive Informatics (ICCI), Beijing, China, 7–9 July 2010; pp. 273–279.

46. Shi, J.-h.; Shen, J.-z.; Ji, Y.; Du, F.-l. A submatrix-based P300 brain-computer interface stimulus presentation paradigm. *J. Zhejiang Univ. Sci. C* **2012**, *13*, 452–459. [CrossRef]

47. Eom, J.S.; Yang, H.R.; Park, M.S.; Sohn, J.H. P300 speller using a new stimulus presentation paradigm. In Proceedings of the 2013 International Winter Workshop on Brain-Computer Interface (BCI), Gangwo, Korea, 18–20 February 2013; pp. 98–99.

48. Jin, J.; Horki, P.; Brunner, C.; Wang, X.; Neuper, C.; Pfurtscheller, G. A new P300 stimulus presentation pattern for EEG-based spelling systems. *Biomed. Tech. (Berl.)* **2010**, *55*, 203–210. [CrossRef] [PubMed]

49. Polprasert, C.; Kukieattikool, P.; Demeechai, T.; Ritcey, J.A.; Siwamogsatham, S. New stimulation pattern design to improve P300-based matrix speller performance at high flash rate. *J. Neural Eng.* **2013**, *10*, 036012. [CrossRef] [PubMed]

50. Kaufmann, T.; Schulz, S.M.; Grunzinger, C.; Kubler, A. Flashing characters with famous faces improves ERP-based brain-computer interface performance. *J. Neural Eng.* **2011**, *8*, 056016. [CrossRef] [PubMed]

51. Li, Q.; Liu, S.; Li, J. Neural Mechanism of P300-Speller Brain-Computer Interface Using Familiar Face Paradigm. In Proceedings of the 2015 International Conference on Network and Information Systems for Computers, Wuhan, China, 23–25 January 2015; pp. 611–614.

52. Li, Q.; Liu, S.; Li, J.; Bai, O. Use of a Green Familiar Faces Paradigm Improves P300-Speller Brain-Computer Interface Performance. *PLoS ONE* **2015**, *10*, e0130325. [CrossRef] [PubMed]

53. Kaufmann, T.; Kubler, A. Beyond maximum speed-a novel two-stimulus paradigm for brain-computer interfaces based on event-related potentials (P300-BCI). *J. Neural Eng.* **2014**, *11*, 056004. [CrossRef] [PubMed]

54. Yeom, S.K.; Fazli, S.; Muller, K.R.; Lee, S.W. An efficient ERP-based brain-computer interface using random set presentation and face familiarity. *PLoS ONE* **2014**, *9*, e111157. [CrossRef] [PubMed]

55. Speier, W.; Deshpande, A.; Cui, L.; Chandravadia, N.; Roberts, D.; Pouratian, N. A comparison of stimulus types in online classification of the P300 speller using language models. *PLoS ONE* **2017**, *12*, e0175382. [CrossRef] [PubMed]

56. Kathner, I.; Kubler, A.; Halder, S. Rapid P300 brain-computer interface communication with a head-mounted display. *Front. Neurosci.* **2015**, *9*, 207. [CrossRef] [PubMed]

57. Ahi, S.T.; Kambara, H.; Koike, Y. A dictionary-driven P300 speller with a modified interface. *IEEE Trans. Neural Syst. Rehabil. Eng.* **2011**, *19*, 6–14. [CrossRef] [PubMed]

58. Li, Y.Q.; Nam, C.S.; Shadden, B.B.; Johnson, S.L. A P300-Based Brain-Computer Interface: Effects of Interface Type and Screen Size. *Int. J. Hum.-Comput. Interact.* **2011**, *27*, 52–68. [CrossRef]

59. Jin, J.; Sellers, E.W.; Wang, X. Targeting an efficient target-to-target interval for P300 speller brain-computer interfaces. *Med. Biol. Eng. Comput.* **2012**, *50*, 289–296. [CrossRef] [PubMed]

60. Sakai, Y.; Yagi, T. Alphabet matrix layout in P300 speller may alter its performance. In Proceedings of the 4th 2011 Biomedical Engineering International Conference, Chiang Mai, Thailand, 29–31 January 2012; pp. 89–92.

61. Speier, W.; Arnold, C.; Pouratian, N. Integrating language models into classifiers for BCI communication: A review. *J. Neural Eng.* **2016**, *13*, 031002. [CrossRef] [PubMed]

62. Ryan, D.B.; Frye, G.E.; Townsend, G.; Berry, D.R.; Mesa, G.S.; Gates, N.A.; Sellers, E.W. Predictive spelling with a P300-based brain-computer interface: Increasing the rate of communication. *Int. J. Hum. Comput. Interact.* **2011**, *27*, 69–84. [CrossRef] [PubMed]

63. Kaufmann, T.; Volker, S.; Gunesch, L.; Kubler, A. Spelling is Just a Click Away—A User-Centered Brain-Computer Interface Including Auto-Calibration and Predictive Text Entry. *Front. Neurosci.* **2012**, *6*, 72. [CrossRef] [PubMed]

64. Akram, F.; Metwally, M.K.; Han, H.S.; Jeon, H.J.; Kim, T.S. A novel P300-based BCI system for words typing. In Proceedings of the 2013 International Winter Workshop on Brain-Computer Interface (BCI), Gangwo, Korea, 18–20 February 2013; pp. 24–25.

65. Akram, F.; Han, H.S.; Kim, T.S. A P300-based brain computer interface system for words typing. *Comput. Biol. Med.* **2014**, *45*, 118–125. [CrossRef] [PubMed]

66. Minett, J.W.; Peng, G.; Zhou, L.; Zheng, H.Y.; Wang, W.S.Y. An Assistive Communication Brain-Computer Interface for Chinese Text Input. In Proceedings of the 2010 4th International Conference on Bioinformatics and Biomedical Engineering, Chengdu, China, 18–20 June 2010; pp. 1–4.

67. Minett, J.W.; Zheng, H.-Y.; Fong, M.C.M.; Zhou, L.; Peng, G.; Wang, W.S.Y. A Chinese Text Input Brain–Computer Interface Based on the P300 Speller. *Int. J. Hum.-Comput. Interact.* **2012**, *28*, 472–483. [CrossRef]

68. Yu, Y.; Zhou, Z.; Yin, E.; Jiang, J.; Liu, Y.; Hu, D. A P300-Based Brain–Computer Interface for Chinese Character Input. *Int. J. Hum.-Comput. Interact.* **2016**, *32*, 878–884. [CrossRef]

69. Kabbara, A.; Hassan, M.; Khalil, M.; Eid, H.; El-Falou, W. An efficient P300-speller for Arabic letters. In Proceedings of the 2015 International Conference on Advances in Biomedical Engineering (ICABME), Beirut, Lebanon, 16–18 September 2015; pp. 142–145.

70. Lee, T.H.; Kam, T.E.; Kim, S.P. A P300-Based Hangul Input System with a Hierarchical Stimulus Presentation Paradigm. In Proceedings of the 2011 International Workshop on Pattern Recognition in NeuroImaging, Seoul, Korea, 16–18 May 2011; pp. 21–24.

71. Yamamoto, Y.; Yoshikawa, T.; Furuhashi, T. Improving performance and accuracy of the P300 speller via a second display. In Proceedings of the 2014 Joint 7th International Conference on Soft Computing and Intelligent Systems (SCIS) and 15th International Symposium on Advanced Intelligent Systems (ISIS), Kitakyushu, Japan, 3–6 December 2014; pp. 793–796.

72. Ikegami, S.; Takano, K.; Kondo, K.; Saeki, N.; Kansaku, K. A region-based two-step P300-based brain-computer interface for patients with amyotrophic lateral sclerosis. *Clin. Neurophysiol.* **2014**, *125*, 2305–2312. [CrossRef] [PubMed]

73. Noorzadeh, S.; Rivet, B.; Jutten, C. Beyond 2D for Brain-Computer interfaces: Two 3D extensions of the P300-Speller. In Proceedings of the 2014 IEEE International Conference on Acoustics, Speech and Signal Processing (ICASSP), Florence, Italy, 4–9 May 2014; pp. 5899–5903.

74. Acqualagna, L.; Treder, M.S.; Blankertz, B. Chroma Speller: Isotropic visual stimuli for truly gaze-independent spelling. In Proceedings of the 2013 6th International IEEE/EMBS Conference on Neural Engineering (NER), San Diego, CA, USA, 6–8 November 2013; pp. 1041–1044.

75. Akram, F.; Han, S.M.; Kim, T.S. An efficient word typing P300-BCI system using a modified T9 interface and random forest classifier. *Comput. Biol. Med.* **2015**, *56*, 30–36. [CrossRef] [PubMed]

76. Ron-Angevin, R.; Varona-Moya, S.; da Silva-Sauer, L. Initial test of a T9-like P300-based speller by an ALS patient. *J. Neural Eng.* **2015**, *12*, 046023. [CrossRef] [PubMed]

77. Postelnicu, C.C.; Talaba, D. P300-based brain-neuronal computer interaction for spelling applications. *IEEE Trans. Biomed. Eng.* **2013**, *60*, 534–543. [CrossRef] [PubMed]

78. Aloise, F.; Arico, P.; Schettini, F.; Riccio, A.; Salinari, S.; Mattia, D.; Babiloni, F.; Cincotti, F. A covert attention P300-based brain-computer interface: Geospell. *Ergonomics* **2012**, *55*, 538–551. [CrossRef] [PubMed]

79. Liu, Y.; Zhou, Z.; Hu, D. Gaze independent brain-computer speller with covert visual search tasks. *Clin. Neurophysiol.* **2011**, *122*, 1127–1136. [CrossRef] [PubMed]

80. Pires, G.; Nunes, U.; Castelo-Branco, M. GIBS block speller: Toward a gaze-independent P300-based BCI. In Proceedings of the 2011 Annual International Conference of the IEEE Engineering in Medicine and Biology Society, Boston, MA, USA, 30 August–3 September 2011; pp. 6360–6364.

81. Pires, G.; Nunes, U.; Castelo-Branco, M. Comparison of a row-column speller vs. a novel lateral single-character speller: Assessment of BCI for severe motor disabled patients. *Clin. Neurophysiol.* **2012**, *123*, 1168–1181. [CrossRef] [PubMed]

82. Zhou, S.; Chen, L.; Jin, J.; Zhang, Y.; Wang, X. Exploring motion visual-evoked potentials for multi-objective gaze-independent brain-computer interfaces. In Proceedings of the 2016 3rd International Conference on Systems and Informatics (ICSAI), Shanghai, China, 19–21 November 2016; pp. 87–91.

83. Treder, M.S.; Schmidt, N.M.; Blankertz, B. Gaze-independent brain-computer interfaces based on covert attention and feature attention. *J. Neural Eng.* **2011**, *8*, 066003. [CrossRef] [PubMed]

84. Schmidt, N.M.; Blankertz, B.; Treder, M.S. Online detection of error-related potentials boosts the performance of mental typewriters. *BMC Neurosci.* **2012**, *13*, 19. [CrossRef] [PubMed]

85. Acqualagna, L.; Treder, M.S.; Schreuder, M.; Blankertz, B. A novel brain-computer interface based on the rapid serial visual presentation paradigm. In Proceedings of the 2010 Annual International Conference of the IEEE Engineering in Medicine and Biology, Buenos Aires, Argentina, 31 August–4 September 2010; pp. 2686–2689.

86. Acqualagna, L.; Blankertz, B. A gaze independent spelling based on rapid serial visual presentation. In Proceedings of the 2011 Annual International Conference of the IEEE Engineering in Medicine and Biology Society, Boston, MA, USA, 30 August–3 September 2011; pp. 4560–4563.

87. Acqualagna, L.; Blankertz, B. Gaze-independent BCI-spelling using rapid serial visual presentation (RSVP). *Clin. Neurophysiol.* **2013**, *124*, 901–908. [CrossRef] [PubMed]

88. Sato, H.; Washizawa, Y. An N100-P300 Spelling Brain-Computer Interface with Detection of Intentional Control. *Computers* **2016**, *5*, 31. [CrossRef]

89. Nakanishi, M.; Wang, Y.; Chen, X.; Wang, Y.T.; Gao, X.; Jung, T.P. Enhancing Detection of SSVEPs for a High-Speed Brain Speller Using Task-Related Component Analysis. *IEEE Trans. Biomed. Eng.* **2018**, *65*, 104–112. [CrossRef] [PubMed]

90. Yin, E.; Zhou, Z.; Jiang, J.; Yu, Y.; Hu, D. A Dynamically Optimized SSVEP Brain-Computer Interface (BCI) Speller. *IEEE Trans. Biomed. Eng.* **2015**, *62*, 1447–1456. [CrossRef] [PubMed]

91. Yin, E.; Zhou, Z.; Jiang, J.; Chen, F.; Liu, Y.; Hu, D. A novel hybrid BCI speller based on the incorporation of SSVEP into the P300 paradigm. *J. Neural Eng.* **2013**, *10*, 026012. [CrossRef] [PubMed]

92. Yin, E.; Zhou, Z.; Jiang, J.; Chen, F.; Liu, Y.; Hu, D. A speedy hybrid BCI spelling approach combining P300 and SSVEP. *IEEE Trans. Biomed. Eng.* **2014**, *61*, 473–483. [CrossRef] [PubMed]

93. Yin, E.; Zeyl, T.; Saab, R.; Chau, T.; Hu, D.; Zhou, Z. A Hybrid Brain-Computer Interface Based on the Fusion of P300 and SSVEP Scores. *IEEE Trans. Neural Syst. Rehabil. Eng.* **2015**, *23*, 693–701. [CrossRef] [PubMed]

94. D'Albis, T.; Blatt, R.; Tedesco, R.; Sbattella, L.; Matteucci, M. A predictive speller controlled by a brain-computer interface based on motor imagery. *ACM Trans. Comput.-Hum. Interact.* **2012**, *19*, 1–25. [CrossRef]

95. Chang, M.H.; Lee, J.S.; Heo, J.; Park, K.S. Eliciting dual-frequency SSVEP using a hybrid SSVEP-P300 BCI. *J. Neurosci. Methods* **2016**, *258*, 104–113. [CrossRef] [PubMed]

96. Lin, K.; Cinetto, A.; Wang, Y.; Chen, X.; Gao, S.; Gao, X. An online hybrid BCI system based on SSVEP and EMG. *J. Neural Eng.* **2016**, *13*, 026020. [CrossRef] [PubMed]

97. Roula, M.A.; Kulon, J.; Mamatjan, Y. Brain-computer interface speller using hybrid P300 and motor imagery signals. In Proceedings of the 2012 4th IEEE RAS & EMBS International Conference on Biomedical Robotics and Biomechatronics (BioRob), Rome, Italy, 24–27 June 2012; pp. 224–227.

98. Kick, C.; Volosyak, I. Evaluation of different spelling layouts for SSVEP based BCIs. In Proceedings of the 2014 36th Annual International Conference of the IEEE Engineering in Medicine and Biology Society, Chicago, IL, USA, 26–30 August 2014; pp. 1634–1637.

99. Fazel-Rezai, R. Human Error in P300 Speller Paradigm for Brain-Computer Interface. In Proceedings of the 2007 29th Annual International Conference of the IEEE Engineering in Medicine and Biology Society, Lyon, France, 22–26 August 2007; pp. 2516–2519.

100. Bian, W.; Yu, S.; Jian-hui, Z.; Xin, L.; Ji-cai, Z.; Wei-dong, C.; Xiao-xiang, Z. A Virtual Chinese Keyboard BCI System Based on P300 Potentials. *Acta Electron. Sin.* **2009**, *37*, 1733–1738.

101. Hohne, J.; Schreuder, M.; Blankertz, B.; Tangermann, M. A Novel 9-Class Auditory ERP Paradigm Driving a Predictive Text Entry System. *Front. Neurosci.* **2011**, *5*, 99. [CrossRef] [PubMed]

102. Townsend, G.; LaPallo, B.K.; Boulay, C.B.; Krusienski, D.J.; Frye, G.E.; Hauser, C.K.; Schwartz, N.E.; Vaughan, T.M.; Wolpaw, J.R.; Sellers, E.W. A novel P300-based brain-computer interface stimulus presentation paradigm: Moving beyond rows and columns. *Clin. Neurophysiol.* **2010**, *121*, 1109–1120. [CrossRef] [PubMed]

103. Arico, P.; Aloise, F.; Schettini, F.; Salinari, S.; Mattia, D.; Cincotti, F. Influence of P300 latency jitter on event related potential-based brain-computer interface performance. *J. Neural Eng.* **2014**, *11*. [CrossRef] [PubMed]

104. Blankertz, B.; Krauledat, M.; Dornhege, G.; Williamson, J.; Murray-Smith, R.; Müller, K.-R. A Note on Brain Actuated Spelling with the Berlin Brain-Computer Interface. *Univers. Access Hum.-Comput. Interact. Ambient Interact.* **2007**, *4555/2007*, 759–768. [CrossRef]

105. Treder, M.S.; Blankertz, B. (C)overt attention and visual speller design in an ERP-based brain-computer interface. *Behav. Brain Funct.* **2010**, *6*, 28. [CrossRef] [PubMed]

106. Volosyak, I.; Cecotti, H.; Valbuena, D.; Graser, A. Evaluation of the Bremen SSVEP based BCI in real world conditions. In Proceedings of the 2009 IEEE International Conference on Rehabilitation Robotics, Kyoto, Japan, 23–26 June 2009; pp. 322–331.

107. Volosyak, I. SSVEP-based Bremen-BCI interface—Boosting information transfer rates. *J. Neural Eng.* **2011**, *8*, 036020. [CrossRef] [PubMed]

108. Nagel, S.; Dreher, W.; Rosenstiel, W.; Spüler, M. The effect of monitor raster latency on VEPs, ERPs and Brain–Computer Interface performance. *J. Neurosci. Methods* **2018**, *295*, 45–50. [CrossRef] [PubMed]

109. Allison, B.Z.; Brunner, C.; Altstatter, C.; Wagner, I.C.; Grissmann, S.; Neuper, C. A hybrid ERD/SSVEP BCI for continuous simultaneous two dimensional cursor control. *J. Neurosci. Methods* **2012**, *209*, 299–307. [CrossRef] [PubMed]

110. Hansen, D.W.; Pece, A. Eye typing off the shelf. In Proceedings of the 2004 IEEE Computer Society Conference on Computer Vision and Pattern Recognition, 2004. CVPR 2004, Washington, DC, USA, 27 June–2 July 2004; Volume 152, pp. II-159–II-164.

111. Stawicki, P.; Gembler, F.; Rezeika, A.; Volosyak, I. A Novel Hybrid Mental Spelling Application Based on Eye Tracking and SSVEP-Based BCI. *Brain Sci.* **2017**, *7*, 35. [CrossRef] [PubMed]

112. Gräser, A.; Heyer, T.; Fotoohi, L.; Lange, U.; Kampe, H.; Enjarini, B.; Heyer, S.; Fragkopoulos, C.; Ristic-Durrant, D. A Supportive FRIEND at Work: Robotic Workplace Assistance for the Disabled. *IEEE Robot. Autom. Mag.* **2013**, *20*, 148–159. [CrossRef]

113. Struijk, L.N. An inductive tongue computer interface for control of computers and assistive devices. *IEEE Trans. Biomed. Eng.* **2006**, *53*, 2594–2597. [CrossRef] [PubMed]

114. Fitzgerald, M.M.; Sposato, B.; Politano, P.; Hetling, J.; O'Neill, W. Comparison of three head-controlled mouse emulators in three light conditions. *Augment. Altern. Commun.* **2009**, *25*, 32–41. [CrossRef] [PubMed]

115. Li, Z.; He, W.; Yang, C.; Qiu, S.; Zhang, L.; Su, C.Y. Teleoperation control of an exoskeleton robot using brain machine interface and visual compressive sensing. In Proceedings of the 2016 12th World Congress on Intelligent Control and Automation (WCICA), Guilin, China, 12–15 June 2016; pp. 1550–1555.

116. Ma, J.; Zhang, Y.; Cichocki, A.; Matsuno, F. A novel EOG/EEG hybrid human-machine interface adopting eye movements and ERPs: Application to robot control. *IEEE Trans. Biomed. Eng.* **2015**, *62*, 876–889. [CrossRef] [PubMed]

117. Zhao, J.; Li, W.; Mao, X.; Hu, H.; Niu, L.; Chen, G. Behavior-Based SSVEP Hierarchical Architecture for Telepresence Control of Humanoid Robot to Achieve Full-Body Movement. *IEEE Trans. Cogn. Dev. Syst.* **2017**, *9*, 197–209. [CrossRef]

118. Lin, Y.T.; Kuo, C.H. Development of SSVEP-based intelligent wheelchair brain computer interface assisted by reactive obstacle avoidance. In Proceedings of the 2016 IEEE International Conference on Industrial Technology (ICIT), Taipei, Taiwan, 14–17 March 2016; pp. 1572–1577.

119. Turnip, M.; Dharma, A.; Pasaribu, H.H.S.; Harahap, M.; Amri, M.F.; Suhendra, M.A.; Turnip, A. An application of online ANFIS classifier for wheelchair based brain computer interface. In Proceedings of the 2015 International Conference on Automation, Cognitive Science, Optics, Micro Electro-Mechanical System, and Information Technology (ICACOMIT), Bandung, Indonesia, 29–30 October 2015; pp. 134–137.

120. Chen, L.; Wang, Z.; Feng, H.; Yang, J.; Qi, H.; Zhou, P.; Wang, B.; Dong, M. An online hybrid brain-computer interface combining multiple physiological signals for webpage browse. In Proceedings of the 2015 37th Annual International Conference of the IEEE Engineering in Medicine and Biology Society (EMBC), Milan, Italy, 25–29 August 2015; pp. 1152–1155. [CrossRef]

121. Spüler, M. A Brain-Computer Interface (BCI) system to use arbitrary Windows applications by directly controlling mouse and keyboard. In Proceedings of the 2015 37th Annual International Conference of the IEEE Engineering in Medicine and Biology Society (EMBC), Milan, Italy, 25–29 August 2015; pp. 1087–1090.

122. Wong, C.M.; Tang, Q.; Cruz, J.N.d.; Wan, F. A multi-channel SSVEP-based BCI for computer games with analogue control. In Proceedings of the 2015 IEEE International Conference on Computational Intelligence and Virtual Environments for Measurement Systems and Applications (CIVEMSA), Shenzhen, China, 12–14 June 2015; pp. 1–6.

123. Schettini, F.; Riccio, A.; Simione, L.; Liberati, G.; Caruso, M.; Frasca, V.; Calabrese, B.; Mecella, M.; Pizzimenti, A.; Inghilleri, M.; et al. Assistive device with conventional, alternative, and brain-computer interface inputs to enhance interaction with the environment for people with amyotrophic lateral sclerosis: A feasibility and usability study. *Arch. Phys. Med. Rehabil.* **2015**, *96*, S46–S53. [CrossRef] [PubMed]

124. Park, H.J.; Kim, K.T.; Lee, S.W. Towards a smart TV control system based on steady-state visual evoked potential. In Proceedings of the 3rd International Winter Conference on Brain-Computer Interface, Sabuk, Korea, 12–14 January 2015; pp. 1–2.

125. Corralejo, R.; Nicolas-Alonso, L.F.; Alvarez, D.; Hornero, R. A P300-based brain-computer interface aimed at operating electronic devices at home for severely disabled people. *Med. Biol. Eng. Comput.* **2014**, *52*, 861–872. [CrossRef] [PubMed]

126. Rohani, D.A.; Puthusserypady, S. BCI inside a virtual reality classroom: A potential training tool for attention. *EPJ Nonlinear Biomed. Phys.* **2015**, *3*, 12. [CrossRef]

127. Volosyak, I.; Valbuena, D.; Malechka, T.; Peuscher, J.; Graser, A. Brain-computer interface using water-based electrodes. *J. Neural Eng.* **2010**, *7*, 066007. [CrossRef] [PubMed]

128. Fei, W.; Guangli, L.; Jingjing, C.; Yanwen, D.; Dan, Z. Novel semi-dry electrodes for brain–computer interface applications. *J. Neural Eng.* **2016**, *13*, 046021.

129. Elsawy, A.S.; Eldawlatly, S.; Taher, M.; Aly, G.M. Enhancement of mobile development of brain-computer platforms. In Proceedings of the 2015 IEEE International Conference on Electronics, Circuits, and Systems (ICECS), Cairo, Egypt, 6–9 December 2015; pp. 490–491.

130. Kaczmarek, P.; Salomon, P. Towards SSVEP-based, portable, responsive Brain-Computer Interface. In Proceedings of the 2015 37th Annual International Conference of the IEEE Engineering in Medicine and Biology Society (EMBC), Milan, Italy, 25–29 August 2015; pp. 1095–1098.

131. Mora, N.; De Munari, I.; Ciampolini, P. SSVEP-based BCI: A "Plug & play" approach. In Proceedings of the 2015 37th Annual International Conference of the IEEE Engineering in Medicine and Biology Society (EMBC), Milan, Italy, 25–29 August 2015; pp. 6170–6173.

132. Gembler, F.; Stawicki, P.; Volosyak, I. Autonomous Parameter Adjustment for SSVEP-Based BCIs with a Novel BCI Wizard. *Front. Neurosci.* **2015**, *9*, 474. [CrossRef] [PubMed]

133. Sengelmann, M.; Engel, A.K.; Maye, A. Maximizing Information Transfer in SSVEP-Based Brain–Computer Interfaces. *IEEE Trans. Biomed. Eng.* **2017**, *64*, 381–394. [CrossRef] [PubMed]

134. Wang, H.; Zhang, Y.; Waytowich, N.R.; Krusienski, D.J.; Zhou, G.; Jin, J.; Wang, X.; Cichocki, A. Discriminative Feature Extraction via Multivariate Linear Regression for SSVEP-Based BCI. *IEEE Trans. Neural Syst. Rehabil. Eng.* **2016**, *24*, 532–541. [CrossRef] [PubMed]

135. Jiao, Y.; Zhang, Y.; Jin, J.; Wang, X. Multilayer correlation maximization for frequency recognition in SSVEP brain-computer interface. In Proceedings of the 2016 Sixth International Conference on Information Science and Technology (ICIST), Dalian, China, 6–8 May 2016; pp. 31–35.

136. Dal Seno, B.; Matteucci, M.; Mainardi, L. Online detection of P300 and error potentials in a BCI speller. *Comput. Intell. Neurosci.* **2010**. [CrossRef] [PubMed]

137. Spüler, M.; Bensch, M.; Kleih, S.; Rosenstiel, W.; Bogdan, M.; Kübler, A. Online use of error-related potentials in healthy users and people with severe motor impairment increases performance of a P300-BCI. *Clin. Neurophysiol.* **2012**, *123*, 1328–1337. [CrossRef] [PubMed]

138. Aloise, F.; Aricò, P.; Schettini, F.; Salinari, S.; Mattia, D.; Cincotti, F. Asynchronous gaze-independent event-related potential-based brain–computer interface. *Artif. Intell. Med.* **2013**, *59*, 61–69. [CrossRef] [PubMed]

139. Schettini, F.; Aloise, F.; Aricò, P.; Salinari, S.; Mattia, D.; Cincotti, F. Self-calibration algorithm in an asynchronous P300-based brain–computer interface. *J. Neural Eng.* **2014**, *11*, 035004. [CrossRef] [PubMed]

140. Belitski, A.; Farquhar, J.; Desain, P. P300 audio-visual speller. *J. Neural Eng.* **2011**, *8*, 025022. [CrossRef] [PubMed]

141. Halder, S.; Käthner, I.; Kübler, A. Training leads to increased auditory brain–computer interface performance of end-users with motor impairments. *Clin. Neurophysiol.* **2016**, *127*, 1288–1296. [CrossRef] [PubMed]

142. Hori, J.; Okada, N. Classification of tactile event-related potential elicited by Braille display for brain–computer interface. *Biocybern. Biomed. Eng.* **2017**, *37*, 135–142. [CrossRef]

143. Yin, E.; Zeyl, T.; Saab, R.; Hu, D.; Zhou, Z.; Chau, T. An auditory-tactile visual saccade-independent P300 brain–computer interface. *Int. J. Neural Syst.* **2016**, *26*, 1650001. [CrossRef] [PubMed]

brain sciences

MDPI

Article

A 20-Questions-Based Binary Spelling Interface for Communication Systems

Alessandro Tonin [1], Niels Birbaumer [1,2] and Ujwal Chaudhary [1,2,*]

[1] Institute of Medical Psychology and Behavioral Neurobiology, University of Tübingen, 72076 Tübingen, Germany; alessandro.tonin@uni-tuebingen.de (A.T.); niels.birbaumer@uni-tuebingen.de (N.B.)

[2] Wyss-Center for Bio- and Neuro-Engineering, 1202 Geneva, Switzerland

* Correspondence: chaudharyujwal@gmail.com; Tel.: +49-707-129-73254

Received: 11 June 2018; Accepted: 30 June 2018; Published: 2 July 2018

Abstract: Brain computer interfaces (BCIs) enables people with motor impairments to communicate using their brain signals by selecting letters and words from a screen. However, these spellers do not work for people in a complete locked-in state (CLIS). For these patients, a near infrared spectroscopy-based BCI has been developed, allowing them to reply to "yes"/"no" questions with a classification accuracy of 70%. Because of the non-optimal accuracy, a usual character-based speller for selecting letters or words cannot be used. In this paper, a novel spelling interface based on the popular 20-questions-game has been presented, which will allow patients to communicate using only "yes"/"no" answers, even in the presence of poor classification accuracy. The communication system is based on an artificial neural network (ANN) that estimates a statement thought by the patient asking less than 20 questions. The ANN has been tested in a web-based version with healthy participants and in offline simulations. Both results indicate that the proposed system can estimate a patient's imagined sentence with an accuracy that varies from 40%, in the case of a "yes"/"no" classification accuracy of 70%, and up to 100% in the best case. These results show that the proposed spelling interface could allow patients in CLIS to express their own thoughts, instead of only answer to "yes"/"no" questions.

Keywords: brain computer interface; complete locked-in state; communication; Artificial Neural Network; 20-questions-game

1. Introduction

In the past decades, many alternative communication systems have been developed for people with speech, language, or motor impairments. Brain computer interfaces (BCI) were developed to provide a means of communication for people with severe motor disabilities (for review see Chaudhary et al., 2016) [1–3]. The most commonly used non-invasive BCI spelling application is based on the electroencephalography (EEG) based P300 event-related brain potential, where a patient can select letters from a matrix in which each character is transiently illuminated [4]. Another BCI system commonly used to select letters from a screen is based on steady state visually evoked potential (SSEVP) [5,6]. Other BCI communication systems are based on slow cortical potential [7], and on the sensorimotor rhythm of the EEG [8,9] to control cursors or keyboards on a screen. These systems, even using different signals and different interfaces, are all based on the same general paradigm, namely, that patients communicate by selecting letters or words from a screen. Different features and classification techniques are used to decode the intention of patients [10–12]. Independently from the signal type, all of these BCI systems are based on the control of a neuroelectric brain response, and the learning process is based on feedback and reward. Despite the good results achievable using these systems with patients suffering from disorders leading to loss of communication, none of these techniques were able to provide a means of communication to amyotrophic lateral sclerosis (ALS) patients in a completely locked-in state (CLIS). An explanation of the non-applicability of the standard BCI in complete paralysis with otherwise intact cognitive processing, Kübler and Birbaumer suggested the theoretical psychophysiological notion of "extinction of goal directed cognition and thought" in CLIS [13]. Following this idea, a BCI based on functional near-infrared spectroscopy (*f*NIRS) was developed for 'reflexive' communication in CLIS. Unlike the other communication systems, it allows the patient to answer short questions affirmatively ("yes") and negatively ("no"), using the blood oxygenation change of their fronto-central brain regions. The best accuracy reported for correctly classified "yes"/"no" answers is 70% in CLIS [14,15]. The low classification accuracy and the only binary "yes"/"no" answers do not allow the patients to express their own thoughts using a classic character-selection-based speller, but only to answer prerecorded questions.

The limitations of the *f*NIRS-BCI, especially the restriction to a binary "yes"/"no" signal and a substantial error rate, are common not only to all non-invasive BCI systems, but also to all the telecommunication systems. Using telecommunication words, the BCI problem involves the correct detection of a communication between two agents through a noisy channel. The communication, both in the general case of telecommunication or in the particular case of the "yes"/"no"-BCI, is a binary message sent from the sender (or the brain) to the receiver (the computer), whose information may be distorted in the transmission due to the noise in the channel (wrong classification), and the task of the receiver is to recover the message reconstructing the corrupted signal [16].

The BCI-spellers usually solve the problem of the wrong signal classification with a redundant number of inputs (e.g., flashing each letter multiple times in order to be sure that the selection was not due to a false positive). With the *f*NIRS-BCI, this technique is because of the characteristic of the *f*NIRS signal; the *f*NIRS-BCI system is slow and allows the patient to answer approximately only one question every 20 s. The solution for this kind of BCI would be a speller capable of correcting the errors in the classification of the answers, allowing a patient to communicate using minimum number of inputs.

A solution can be found in a popular game, the 20-questions-game. In this game, a player has to guess what the other player is thinking within 20 "yes"/"no" questions. An electronic version of the game, which has been played more than 88 millions times, can correctly guess what someone is thinking with 80% precision, by asking 20 questions (95% of the time with 25 questions) [17]. The game was mathematically formalized by Alfred Rényi [18] and it was later proposed in a different version by Stanisław Ulam [19]. The Rényi¬–Ulam game and its variations have been used to solve many different problems [20–22], in this paper we propose to use the game as a spelling interface for a binary BCI, like the *f*NIRS-based BCI described in Chaudhary et al. (2017). This kind of communication system may allow patients in CLIS to express their own thoughts and not just to reply to prerecorded questions.

The rest of the paper is structured as follows: in Section 2, the method used to design the communication system is described, and in particular, in Sections 2.1 and 2.2 describe the algorithm of the Rényi–Ulam game and its application to the popular 20-questions-game using an artificial neural network, and in Section 2.3, the implementation as an interface for a BCI system is described. In Section 3, the proposed algorithm is explained in detail. Then, in Section 4, we present the results of the algorithm, both for an online version of the game played by real persons (Section 4.1) and for an offline version with computer simulations (Section 4.2). The results are discussed and followed by the conclusion in Section 5. While the databases used for the results are described in Appendix A.

2. Materials and Methods

2.1. Rényi–Ulam Game

The 20-questions-game is a popular game played by two players. The rules of the game are as follows: the first player (player *A*, the Responder) imagines a famous person, while the second (player *B*, the Questioner) must guess the person by asking twenty "yes"/"no" questions (e.g., "Is the person alive?").

The game has been mathematically described by Rényi and Ulam, as follows: the Responder can imagine any target statement that is contained in a fixed search space (i.e., the topic, e.g., famous people), while the Questioner has to guess the statement using less than n (e.g., 20) "yes"/"no" questions. Moreover, the Responder is allowed to lie up to e times on the answers given to the "yes"/"no" questions (i.e., they can give wrong answers). The lies are a formalization of the wrong answer that a player can give if their knowledge about the statement is different from the knowledge of the other player (e.g., the Responder thinks that a person is alive, but instead it is dead).

The complete description of the game is outlined below:

1. The game is played by two players: A (the Responder) and B (the Questioner).
2. A set S of target statements (the search space) is fixed.
3. A number $n > 0$ of questions is fixed.
4. An upper bound $e \geq 0$ of number of lies is fixed.
5. B can ask questions in the form of "Is x in T?", where T is a subset of S.
6. A must reply "yes" or "no", and he can lie up to e times.
7. B wins if he can correctly guess x after n questions.

The number of questions n to solve the Rényi–Ulam game depends linearly on the cardinality of S and on the maximum number of lies e, but for the general case of an arbitrary number of lies, there is no general solution and only heuristic methods have been proposed [23].

2.2. Artificial Neural Network

A heuristic solution of the Rényi–Ulam game with arbitrary number of lies can be found using an artificial neural network (ANN). This method was first developed by Robin Burgener [24] for *20q*, an electronic version of the 20-question-game. This version is slightly different from the Rényi–Ulam game; for instance, the allowed answers are not only "yes" and "no", but also "unknown", "irrelevant", "sometimes", "depends", etc. Here, we propose an ANN for the original Rényi–Ulam game with binary answers only.

The ANN will play the role of the Questioner, that is, it will ask questions, and it will estimate a particular target statement (e.g., a person) imagined by a Responder. Therefore, in order to work, the ANN needs two databases, one with the target statements belonging to the search space (e.g., all of the possible famous people), and one with the possible "yes"/"no" questions (e.g., "Is it alive?", "Is it a woman?", etc.).

The main core of the ANN is the relation between the statements and questions. Each target statement is connected to each question, and the strength of this connection is indicated by a weight. The weights can be negative if the statement and question are not related (i.e., the expected answer is "no") and positive if they are related (i.e., the expected answer is "yes"). All of the weights are stored in a matrix called a weight matrix.

The ANN will present to the Responder the questions stored in the database. The choice of the question is based on the weight table and on the previous questions.

The final estimation of the ANN is the statement that, based on the received answers, is the most probable. In order to calculate this probability, after each question, the ANN will penalize or reward, based on the answer, the target statements (e.g., if the answer to "Is she a woman?" is "yes", all male persons will be penalized).

Finally, after each correct final estimation, the weight matrix is updated based on the received answer, allowing a learning process.

Using ANN has two advantages. First, if the Responder occasionally lies, the ANN will not exclude any possible target statement, based on that single answer, but it will only change the probability for the final estimation. Second, the estimation of the target statement will improve with frequent usage of ANN, because the learning process improves the reliability of the weight table.

2.3. 20-Questions-Based Interface for Communication Systems

2.3.1. Proposed BCI Implementation

We endeavor to use the 20-questions-game as a communication system for patients that do not have a reliable means of communication, like patients in a complete locked-in state (CLIS). This system is based on an ANN that interacts with the patient in a 20-questions-based paradigm, in order to estimate their thoughts.

For this purpose, the ANN can be developed as part of a brain-computer interface; the computer proposes auditorily the questions to the patient, and it records a brain signal (e.g., *f*NIRS). The BCI classifies the brain signal in a binary answer ("yes" or "no"), which will be the answer required by the ANN. In this implementation, the patient will play the role of the Responder, while the ANN will be the Questioner. The patient can think of any word or sentence that is stored in the database of the ANN, and the ANN will ask questions, also stored in the database, in order to estimate the patient's thought. The "yes"/"no" classification accuracy achieved using BCI systems with CLIS patients is around 70% [14,15]. Using the 20-questions-based system, the errors on the "yes"/"no" classification will be considered as the lies of the Rényi–Ulam game, therefore, they will not automatically lead to a wrong estimation of the sentence.

The proposed 20-questions-based communication system is depicted in Figure 1. The system has been tested as a communication system, independently from the brain signal records, with healthy participants, using a web interface, and with computer simulations.

2.3.2. Web-Based Implementation

The web-based version of the algorithm (www.alsbci.eu) was written in Python and it has been translated into three languages, English, German, and Italian.

In the website, the user is asked to put himself in a complete locked-in patient's shoes, playing the 20-questions-game by thinking a sentence that could be asked by a patient in such conditions. The search space was intentionally left ambiguous and not bound to a specific topic, in order to check the performance of the system in a not optimal scenario. The user had also the option to check the list of target statements already stored in the database.

During the game, the ANN presented the questions to the user, who had the opportunity to reply "yes", "no", or "unsure". In the case of an "unsure" answer, the ANN ignored the answer and, instead, it was asking a different question. At the end of each game, the ANN tried to estimate the thought sentence three times, proposing to the users the three most probable targets (i.e., the three statements with the highest current value). Finally, if none of the proposed target statements was the correct one, the user could select (or, if not present, insert) the thought sentence directly from the database.

From the website, the users had also the opportunity to improve the databases of the ANN by adding new statements and questions.

The web-based version was initialized with an initial database manually populated with a set of 41 target statements and 25 questions. The website has been online, accessible to everyone since November 2017. Since then, the game has been played 92 times, and 50 new statements and 113 new questions have been added to the system, bringing the total number to 91 statements and 138 questions, respectively (see Appendix A).

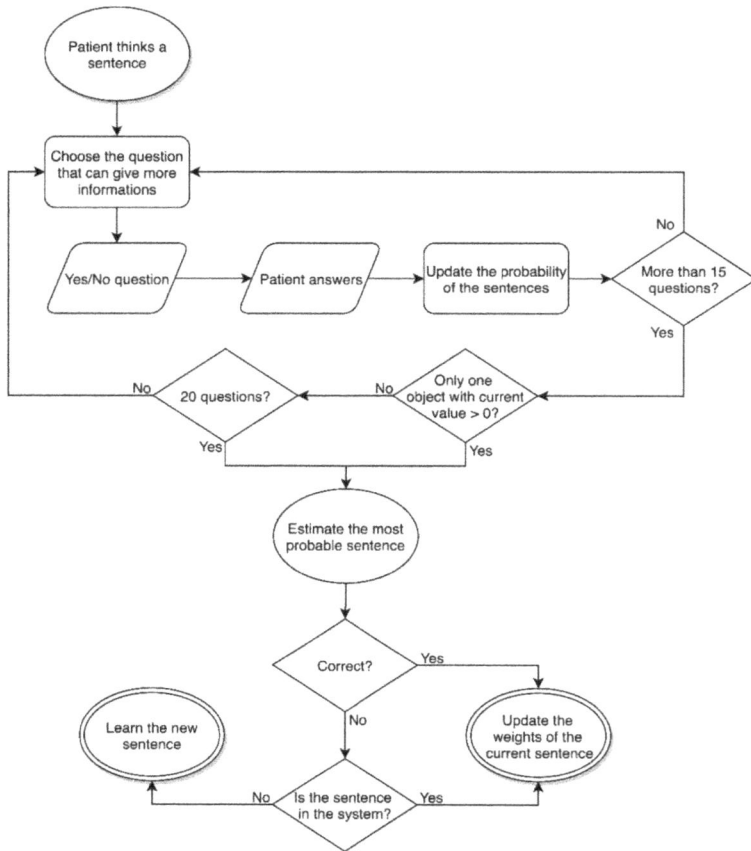

Figure 1. Flow chart of the proposed 20-questions-based communication system.

2.3.3. Simulation

Using an offline version of the website, we tested the algorithm by changing the possible answers and simulating a BCI with errors on the classification of the "yes" and "no" answers.

Regarding the possible answers, we considered three different cases, as follows:

1. "yes", "no", and "unsure" answers, with the questions answered as "unsure" excluded from the total number of questions (same as the online system);
2. "yes", "no", and "unsure" answers, with the questions answered as "unsure" included in the total number of questions; and
3. "yes" and "no" answers only.

As the expected answer is a direct expression of the target-question weight, we considered a "yes" answer when the weight was positive, "no" when negative, and "unsure" when the weight was zero. In the third case, considering the "yes" and "no" answers only, if the target-question weight was zero, we chose "yes" or "no" randomly.

In order to emulate the non-optimal BCI classification, according to the simulated accuracy, each answer had a certain probability of being wrong (if "unsure", the answer was not changed). The algorithm performance has been tested, varying the classification accuracy between 50%

(i.e., random classification) and 100% (i.e., perfect classification). As for the online and the offline analyses, we considered a statement as correctly estimated if, after 20 questions, it was among the three most probable target statements.

3. Algorithm

3.1. Definitions

The two main agents of the ANN are the target statements (i.e., the possible final sentences) and the questions (i.e., the descriptors of the sentences). Both of the target statements and sentences are stored in a database, therefore, the only possible sentences and questions are the ones present in the communication system.

As explained in Section 2.2, the core of the ANN is the weight matrix that puts in relation the target statements and questions. The weight depends on the answer that each statement is required from each question (i.e., if the expected answer is "yes", the weight will be positive, if "no", it will be negative).

A value is assigned to each statement. This value indicates the probability of each statement to be the final target; the higher the value assigned to one statement, the higher the probability of that statement to be the thought one. The value is updated after each question, based on the statement–question weight and on the received answer.

The elements of the ANN are shown in Figure 2, and are summarized below:

- N targets (T_i with $i = 1:N$) (i.e., sentences thought by the patient);
- Each target is described by M descriptors (D_j with $j = 1:M$) (i.e., "yes"/"no" questions);
- Strength of T–D connection is expressed by a weight ($W_{Ti,Dj}$ with $i = 1:N, j = 1:M$); and
- Each target T_i is ranked using a current value (V_{Ti} with $i = 1:N$).

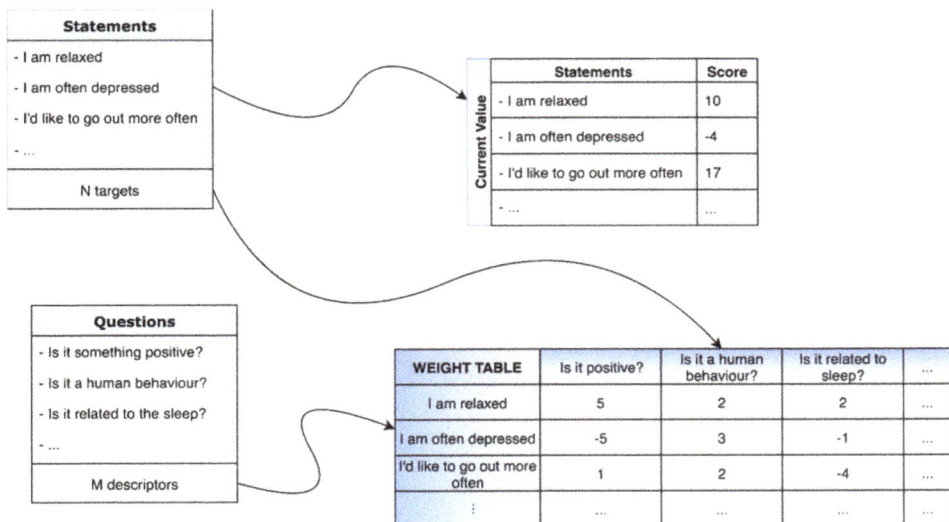

Figure 2. Structure of the artificial neural network. In particular, the structure of the databases of statements and questions, of the table of current values, and of the weight table are shown.

3.2. Current Value Adjustment

During each run, all of the target statements start with the same probability of being the final sentence, therefore, all of the current values V_T are initialized to 0. This probability (i.e., the current value) changes after each presented question, based on the answer of the user. In particular, if D_j is the n-th question presented to the user, for each target statement T_i, the current value V_{T_i} is updated using the formula, as follows:

$$V_{T_i}(n) = V_{T_i}(n-1) + W_{T_i,D_j} \text{ if answer is "yes"}$$

$$V_{T_i}(n) = V_{T_i}(n-1) - W_{T_i, D_j} \text{ if answer is "no"}$$

where n is the number of the question, and W_{T_i,D_j} is the weight between question D_j and statement T_i. It is positive if the expected answer is "yes" and negative if the expected answer is "no". Therefore, the formula increases the current value if the given answer is the expected one, and decreases it otherwise.

In order to decrease the impact of the wrong answers, the adjustment of the current value has been increased for those statements that receive many answers coherent with the expected ones. After each question, every statement where the expected answer matches with the received one is marked as a 'priority target'. This priority is lost whenever the statement receives an answer that does not match with the expected answer. The priority targets receive an adjustment for their current value, equal to double the weight. This leads to the following modified formula for updating the current value:

$$V_{T_i}(n) = V_{T_i}(n-1) + W_{T_i,D_j}(\times 2 \text{ if } T_i \text{ has priority}) \text{ if answer is "yes"}$$

$$V_{T_i}(n) = V_{T_i}(n-1) - W_{T_i, D_j}(\times 2 \text{ if } T_i \text{ has priority}) \text{ if answer is "no"}$$

where the variables are the same as described above.

3.3. Choice of the Question

One of the crucial points of the algorithm is the choice of the question. The best question is the one whose answer will give more information about the most probable targets, or, in other words, the one whose answer splits the most probable targets in two similar sets. Therefore, the best question is the one that maximizes the entropy

$$H(D_j) = \sum_{x \in X} -p(x) \log_2 p(x)$$

where X is the two classes of statements with positive and negative weights, with respect to the question D_j; and $p(x)$ is the proportion of the most probable statements that belong to the class x.

In the implementation, all of the targets with a positive current value were considered as the most probable targets. It is possible to choose the most probable targets in a different way, using a more or less strict definition (e.g., the targets with a current value greater than a certain threshold), and this will obviously change the choice of the questions accordingly.

3.4. Estimate the Target

The goal of the ANN is to estimate the target statement that the patient is thinking. After 15 questions, the ANN will check if there is only one target statement with a positive value; if this happens, it will estimate that statement. If this condition never occurs, after 20 questions, the ANN will estimate the target statement with the highest current value.

The lower threshold of 15 questions is based on the minimum number of questions needed for an optimal solution of the Rényi–Ulam game; considering a search space of 91 statements and a signal classification accuracy of 75%, the minimum number of questions for a deterministic optimal solution is 23 (Table 2.3 from Cicalese, 2013, p. 28). We decided to check whether there was only one statement with a positive value after two thirds of the minimum number of questions for an optimal solution. This condition is meant to speed up the communication process, avoiding asking unnecessary questions when one statement is likely the correct target.

3.5. Learning Step

The last step of the algorithm is teaching the neural network. After each correct estimation, the system will update the weight matrix. For each question that was asked during the run, it will update the weight that associates that question to the correctly estimated statement, based on the answer that the user gave; if the given answer is "yes", it will increase the weight value, otherwise it will decrease it. In order to avoid excessive values, the weights are upper and lower bounded.

4. Results

In the next paragraphs the online and offline results of the proposed algorithm will be presented. The results are based on the web-based version and on the simulations descripted in Sections 2.3.2 and 2.3.3, respectively.

4.1. Online Results

The results of the games played online are reported in Table 1. Half of the time the game was played with a statement that was not in the system; considering that only the games that played with statements already in the system, the percentage of correct estimations is 65.95%, against 34.04% of games where the ANN was not able to correctly estimate the thought sentence. Focusing on the sentences correctly estimated, 67.74% of the time the sentence was estimated on the first attempt.

Table 1. Results of the game played online on the website. The table lists the total number of times of the game play. The game was played for a total of 92 times, out of which it was played for 45 times on new statements (not in the database) and 47 times on old statements (in the database). For the statements already in the database, the table also lists the number of times that they were estimated incorrectly and correctly. For the correctly estimated statements the table lists the number of times the statements were the first, second, or third guess.

New Statements	Old Statements		
45	47		
	Incorrect	Correct	
	16	31	
	1st Estimation	2nd Estimation	3rd Estimation
	21	5	5

4.2. Offline Results

The offline results, reported in Figure 3, were obtained by simulating the performance of the ANN in the cases mentioned in Section 2.3.3. For each of the three cases, the simulation was performed by varying the signal classification accuracy between random (i.e., 50%) and perfect (i.e., 100%). Figure 3a–c represents the percentage of statements correctly estimated by the ANN after 1000 simulations, with respect to the simulated BCI classification accuracy of "yes" and "no". In each figure, blue, green, and yellow represent the percentage of statements correctly estimated as the most, second most, and third most probable statement, respectively.

In order to evaluate the time performance of the proposed communication system, we compared the typing speed of the ANN to those of the classic P300-based matrix speller [25]. The *f*NIRS-based BCI developed for CLIS patients is able to present one question every 20 s [15]. Therefore, a spelling interface that uses this BCI has an information transfer rate (ITR) of 3 bits/min, while the matrix speller reaches 12 bits/min, which means a typing speed of approximately one character every 26 s. The target statements in the database of the ANN (Table A1) have an average length of 23.625 characters. Hence, as in the simulations, the statements were estimated in 20 questions, the *f*NIRS-BCI for the CLIS patients using the 20-questions-based spelling interface will have an average typing speed of one character every 17 s.

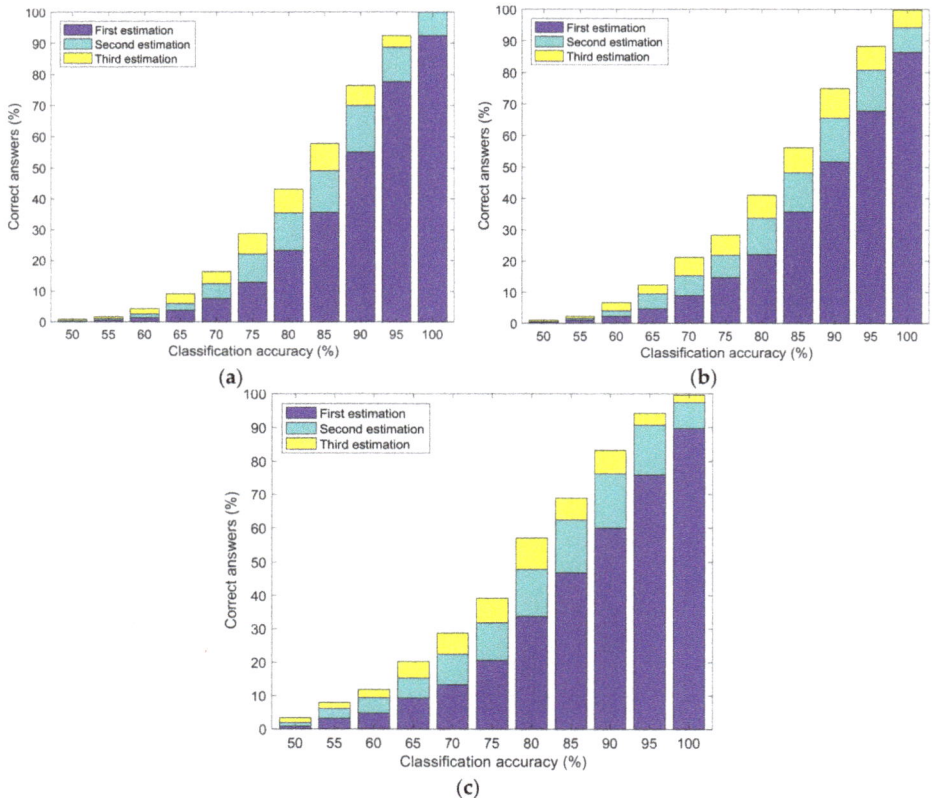

Figure 3. Results of the offline simulation in the three different cases. Blue, green, and yellow represent the percentage of statements correctly estimated as most, second most, and third most probable statement, respectively. (**a**) Simulated results using "yes", "no", and "unsure" answers, with the questions answered as "unsure" excluded from the total number of questions; (**b**) simulated results using "yes", "no", and "unsure" answers, with the questions answered as "unsure" included in the total number of questions; and (**c**) simulated results using "yes" and "no" answers only.

5. Discussion and Conclusions

The results in the offline analyses show that the performances are very similar in the first two analyzed cases, discarding and including "unsure" answers. Surprisingly, when giving random answers instead of "unsure", the results improve. We believe that this is due to the randomization of the target statements and does not represent a real improvement in the results.

Figure 3a–c shows that considering a classification accuracy of 100%, the ANN is always able to correctly estimate the target statement. This result means that, using a BCI that perfectly classifies "yes" and "no" answers, a patient could communicate entire words, or even sentences, by answering only 20 questions. The result is very promising, considering that, under the condition of a perfect signal classification, in order to select one character, a usual 6 × 6 grid-based speller needs at least 12 inputs [26].

However, we also notice that if the accuracy drops down to 80%, the correct rate decreases to 57%. Nevertheless, we have to consider that we did not put any constraint on the possible target statements, so in the same database, there were very different sentences like, "This movie is beautiful" and "I would like to go more out from the bed". This generality of the sentences put the program in a bad case scenario. Although, it is important to notice that these results are still significant, as, considering a random classification (accuracy of 50%), the correct rate is close to 0%.

Both in the online games and in the simulations, the system always asked 20 questions, therefore, after 15 questions, there were always at least two statements with positive value. Hence, the ANN always estimated the final target statement with a certain degree of uncertainty, probably because the number of played games was not enough for an optimal training of the weight table. In order to decrease the uncertainty, a possibility is to increase the number of questions from 20 to the optimal solution number, which depends on the cardinality of the search space and on the signal classification accuracy, as shown in Table 2.3, from Cicalese, 2013, p. 28. Nonetheless, we decided to keep the upper limit of 20 questions in order to build a communication system that could be used in a reasonable time, even using a fNIRS-based BCI (20 s for each question).

The comparison between the 20-questions-based system and the P300 matrix speller shows that, despite a lower ITR, the average typing speed of the proposed spelling interface is higher. Even if this result cannot be taken as a real typing speed comparison because the ANN can estimate only entire sentences, it shows that the proposed system has time performance comparable to the usual spellers and could allow communication in a reasonable time, even in presence of a slow signal like the fNIRS (3 bits/min).

Correlating the online and the offline results, we can say that the users gave the expected answers up to 85% of the time. Obviously, in that case, there were no errors in the signal classification, but we could not expect a perfect result because the questions could have been very general, and with a not unique answer (e.g., considering the sentence "I sleep a lot", the question "Is it positive?" could be answered "yes" or "no" depending on the positive or negative connotation that a person gives to sleeping a lot).

The results show that the 20-questions-based system can be a valid interface for any BCI that uses a slow signal and/or has a classification with a low accuracy rate. Even in presence of fast signal (e.g., EEG), the proposed system can improve the typing speed performance, allowing the formulation of entire sentences using only 20 binary inputs. The main drawback, already highlighted in the previous sections, is that the only sentences that the ANN can estimate are the ones stored in the database, therefore, a patient will not be free to formulate his own sentences. This limitation, an intrinsic characteristic of a 20-questions-system, can be overcome by building an exhaustive database personalized for each patient. Before initiating any BCI session, the patients will be provided an option to choose between the proposed 20-questions-based system and a character-selection speller that gives more freedom at the expense of the typing speed and the error handling.

In the future, we will test the system by narrowing the possible sentences to a more restricted topic and personalizing the weight table for only one person, in order to adapt the weights to his or her individual biography and personality. Moreover, the system will be improved to work with multi-class BCIs, in order to have more possible answers and, therefore, better estimations. Finally, the interface will be tested with a BCI to study the reaction of the patients to this different approach of communication.

The results are promising and show that a communication system based on this algorithm could replace the usual speller-based approach. The main limitation of the 20-questions-based interface is that it does not allow the patient to create new sentences or new questions. Nevertheless, it could allow patients in CLIS to express their own thoughts and desires, instead of only answering to "yes"/"no" questions chosen by someone else. For this reason, the communication system based on the proposed algorithm could be applied to estimate the inner mental and thought process of patients in CLIS.

Author Contributions: Conceptualization, A.T. and U.C.; Methodology, A.T., N.B. and U.C.; Software, A.T.; Validation, A.T., N.B. and U.C.; Formal Analysis, A.T.; Investigation, A.T., N.B. and U.C.; Resources, N.B. and U.C.; Data Curation, A.T. and U.C.; Writing-Original Draft Preparation, A.T.; Writing-Review & Editing, N.B. and U.C.; Visualization, A.T., N.B. and U.C.; Supervision, N.B. and U.C.; Project Administration, N.B. and U.C.; Funding Acquisition, N.B. and U.C.

Funding: Deutsche Forschungsgemeinschaft (DFG, BI 195/77-1), BMBF (German Ministry of Education and Research) 16SV7701 CoMiCon, Baden-Württemberg Stiftung, LUMINOUS-H2020- FETOPEN-2014- 2015-RIA (686764), and Wyss Center for Bio and Neuroengineering, Geneva.

Conflicts of Interest: The authors declare no conflict of interest.

Appendix

Reported here is the complete list of the statements and questions used both in the online and offline results.

Table A1. List of statements and questions used for the online 20-questions-system and offline simulations.

Statements		Questions	
"I am pleased with life"	"I want to travel"	"Would you like to be killed?"	"Is it related to a particular time of day?"
"I am living with pleasure"	"I love my brothers"	"Are you suffering?"	"Is it related to a means of transport?"
"I feel good right now"	"I want to sleep"	"Are you happy with your life?"	"Is it pleasant?"
"I feel bad right now"	"I am thirsty"	"Should I bring you something?"	"Is it just something about fantasy?"
"Most of the time I feel good"	"How beautiful is this movie!"	"Is it something about everyday life?"	"Is it intriguing?"
"Most of the time I feel bad"	"I want to know what the weather will be tomorrow"	"Is it about someone you know?"	"Is it funny?"
"I sleep mostly good"	"I want a beer"	"Is it a daily human need?"	"Is it fun?"
"I sleep mostly bad"	"I love my child"	"It involves a difficult test?"	"Is it exciting?"
"I sleep a lot"	"I would like to go on holiday in Sardinia"	"It has to do with the sea?"	"Is it an entertainment activity?"
"I sleep less"	"I would like to win scientific recognition"	"Is it a desire?"	"Is it an activity that can be associated with routine?"
"I also sleep during the day"	"I want an orange juice"	"Is this something that needs to be cooked?"	"Is it about your hygiene?"
"I sleep only in the night"	"I want to play the guitar"	"Is this something about your career?"	"Is it about the weather?"
"I can concentrate myself on questions"	"I want to have a shower"	"Is there anyone able to do the imagined action?"	"Is it about the future?"
"I cannot concentrate myself on questions"	"I am happy"	"Is the desire for enjoyment?"	"Is it about the bed?"
"I would like to go more out from the bed"	"The music"	"Is it something you do before you sleep?"	"Is it about sex?"
"I like to stay in bed"	"I want to read the newspaper"	"Is it something that you want to do often?"	"Is it about meeting your dreams?"
"I feel very relaxed"	"I had a nice dream"	"Is it something that you do in your house?"	"Is it about human needs?"
"I feel very stressed"	"Some people are really idiots"	"Is it something that you can do without?"	"Is it about food?"
"I am stressed"	"I am stupid"	"Is it something related to a specific season?"	"Is it about an animal?"

Table A1. *Cont.*

Statements		Questions	
"I am relaxed"	"I want to drink a coffee"	"Is it the result of hard work?"	"Is it a wish?"
"I would like to have more visitors"	"I want to play football"	"Is it something you want to do now?"	"Is it a pastime?"
"I would like to have less visitors"	"I wish the best for my loved ones"	"Is it something you eat?"	"Is it a human behavior?"
"I wish more rest"	"I want to go to the gym"	"Is it something to do indoor?"	"Is it a feeling?"
"I am glad when someone visits me"	"My cats are beautiful"	"Is it something to do in the open air?"	"Does it open your mind?"
"My life is good"	"I want to go boating"	"Is it something to do alone?"	"Does it need many attempts and failures?"
"My life is bad"	"I want to eat chocolate"	"Is it something to do accompanied?"	"Does it involve taking revenge?"
"I imagine I am walking"	"I would like to go out more often"	"Is it something that makes you happy?"	"Does it imply a shift?"
"I imagine I am running"	"I am rarely depressed"	"Is it something related to your city?"	"Does it have two eyes?"
"I imagine often I am flying"	"I am often depressed"	"Is it something regarding your loved ones?"	"Does it have to do with music?"
"I imagine often I am eating"	"I laugh often inside myself"	"Is it something positive?"	"Does it have something to do with drinking?"
"I dream a lot"	"I laugh rarely inside myself"	"Is it something physical?"	"Does it have something to do with a candy?"
"I dream less"	"I am hungry"	"Is it something negative?"	"Does it have anything to do with you?"
"I often think soon I will get better"	"I want a cat"	"Is it something emotional?"	"Does it concern your feelings?"
"Rarely I think I will get better soon"	"I want to have sex"	"Is it something abstract?"	"Does it concern nature?"
"I would like it if ... will be more often by me"	"I like to ride a bike"	"Is it something about your family?"	"Does it concern an anatomical part of a person?"
"I am glad that ... is by me"	"I am sleepy"	"Is it something about the sense of hearing?"	"Do you think about it often?"
		"Is it something about the drinks?"	"Do you need company?"
		"Is it something about being free?"	"Do you need an instrument?"
		"Is it something about a primary need?"	"Do you need a ball?"
		"Is it related with the body (care, etc.)?"	"Do you have a need?"
		"Is it related to the present"	"Do you do it for being in the company?"
		"Is it related to the night?"	"Do you do it either alone or in company?"
		"Is it related to the day?"	"Do you do because you need it?"
		"Is it related to sleep?"	"Can you do it alone?"
		"Is it related to imagination?"	"Are you sleepy?"
		"Is it related to a sport?"	"A tool is needed?"

References

1. Chaudhary, U.; Birbaumer, N.; Curado, M.R. Brain-Machine Interface (BMI) in paralysis. *Ann. Phys. Rehabil. Med.* **2015**, *58*, 9–13. [CrossRef] [PubMed]

2. Chaudhary, U.; Birbaumer, N.; Ramos-Murguialday, A. Brain–computer interfaces in the completely locked-in state and chronic stroke. *Prog. Brain Res.* **2016**, *228*, 131–161. [PubMed]

3. Chaudhary, U.; Birbaumer, N.; Ramos-Murguialday, A. Brain-computer interfaces for communication and rehabilitation. *Nat. Rev. Neurol.* **2016**, *12*, 513–525. [CrossRef] [PubMed]

4. Kübler, A.; Furdea, A.; Halder, S.; Hammer, E.M.; Nijboer, F.; Kotchoubey, B. A brain-computer interface controlled auditory event-related potential (p300) spelling system for locked-in patients. *Ann. N. Y. Acad. Sci.* **2009**, *1157*, 90–100. [CrossRef] [PubMed]

5. Volosyak, I. SSVEP-based Bremen-BCI interface—Boosting information transfer rates. *J. Neural Eng.* **2011**, *8*, 036020. [CrossRef] [PubMed]

6. Jiao, Y.; Zhang, Y.; Wang, Y.; Wang, B.; Jin, J.; Wang, X. A novel multilayer correlation maximization model for improving CCA-based frequency recognition in SSVEP brain—Computer interface. *Int. J. Neural Syst.* **2018**, *28*, 1750039. [CrossRef] [PubMed]

7. Neumann, N.; Hinterberger, T.; Kaiser, J.; Leins, U.; Birbaumer, N.; Kübler, A. Automatic processing of self-regulation of slow cortical potentials: Evidence from brain-computer communication in paralysed patients. *Clin. Neurophysiol.* **2004**, *115*, 628–635. [CrossRef] [PubMed]

8. Kübler, A.; Nijboer, F.; Mellinger, J.; Vaughan, T.M.; Pawelzik, H.; Schalk, G.; McFarland, D.J.; Birbaumer, N.; Wolpaw, J.R. Patients with ALS can use sensorimotor rhythms to operate a brain-computer interface. *Neurology* **2005**, *64*, 1775–1777. [CrossRef] [PubMed]

9. Yang, Y.; Chevallier, S.; Wiart, J.; Bloch, I. Subject-specific time-frequency selection for multi-class motor imagery-based BCIs using few Laplacian EEG channels. *Biomed. Signal Process. Control* **2017**, *38*, 302–311. [CrossRef]

10. Zhang, Y.; Zhou, G.; Jin, J.; Zhao, Q.; Wang, X.; Cichocki, A. Sparse Bayesian Classification of EEG for Brain-Computer Interface. *IEEE Trans. Neural Netw. Learn. Syst.* **2015**, *27*, 2256–2267. [CrossRef] [PubMed]

11. Jiao, Y.; Zhang, Y.; Chen, X.; Yin, E.; Jin, J.; Wang, X.Y.; Cichocki, A. Sparse Group Representation Model for Motor Imagery EEG Classification. *IEEE J. Biomed. Heal. Inform.* **2018**. [CrossRef]

12. Zhang, Y.; Wang, Y.; Zhou, G.; Jin, J.; Wang, B.; Wang, X.; Cichocki, A. Multi-kernel extreme learning machine for EEG classification in brain-computer interfaces. *Expert Syst. Appl.* **2018**, *96*, 302–310. [CrossRef]

13. Kübler, A.; Birbaumer, N. Brain-computer interfaces and communication in paralysis: Extinction of goal directed thinking in completely paralysed patients? *Clin. Neurophysiol.* **2008**, *119*, 2658–2666. [CrossRef] [PubMed]

14. Gallegos-Ayala, G.; Furdea, A.; Takano, K.; Ruf, C.A.; Flor, H.; Birbaumer, N. Brain communication in a completely locked-in patient using bedside near-infrared spectroscopy. *Neurology* **2014**, *82*, 1930–1932. [CrossRef] [PubMed]

15. Chaudhary, U.; Xia, B.; Silvoni, S.; Cohen, L.G.; Birbaumer, N. Brain–Computer Interface–Based Communication in the Completely Locked-In State. *PLoS Biol.* **2017**, *15*. [CrossRef] [PubMed]

16. Pelc, A. Searching games with errors—Fifty years of coping with liars. *Theor. Comput. Sci.* **2002**, *270*, 71–109. [CrossRef]

17. 20q.net. Available online: www.webcitation.org/70btN1OFO (accessed on 2 July 2018).

18. Rényi, A. On a problem of information theory. *MTA Mat. Kut. Int. Kozl.* **1961**, *6B*, 505–516.

19. Ulam, S.M. *Adventures of a Mathematician*; University of California Press: Berkeley, CA, USA, 1991; p. 384. ISBN 9780520071544.

20. Jedynak, B.; Frazier, P.I.; Sznitman, R. Twenty questions with noise: Bayes optimal policies for entropy loss. *J. Appl. Probab.* **2012**, *49*, 114–136. [CrossRef]

21. Kazemzadeh, A.; Lee, S.; Georgiou, P.G.; Narayanan, S.S. Emotion twenty questions: Toward a crowd-sourced theory of emotions. In Proceedings of the International Conference on Affective Computing and Intelligent Interaction, Memphis, TN, USA, 9–12 October 2011; Volume 6975 LNCS, pp. 1–10. [CrossRef]

22. Tsiligkaridis, T.; Sadler, B.M.; Hero, A.O. Collaborative 20 Questions for Target Localization. *IEEE Trans. Inf. Theory* **2014**, *60*, 2233–2252. [CrossRef]

23. Cicalese, F. *Fault-Tolerant Search Algorithms*; Monographs in Theoretical Computer Science; An EATCS Series; Springer: Berlin/Heidelberg, Germany, 2013; ISBN 9783642173264.
24. Burgener, R. Artificial Neural Network Guessing Method and Game. U.S. Patent 2010/0311130 Al, 12 October 2006.
25. Farwell, L.A.; Donchin, E. Talking off the top of your head: Toward a mental prosthesis utilizing event-related brain potentials. *Electroencephalogr. Clin. Neurophysiol.* **1988**, *70*, 510–523. [CrossRef]
26. Rezeika, A.; Benda, M.; Stawicki, P.; Gembler, F.; Saboor, A.; Volosyak, I. Brain–Computer Interface Spellers: A Review. *Brain Sci.* **2018**, *8*, 57. [CrossRef] [PubMed]

brain
sciences

MDPI

Article

EEG Waveform Analysis of P300 ERP with Applications to Brain Computer Interfaces

Rodrigo Ramele *, Ana Julia Villar and Juan Miguel Santos †

Computer Engineering Department, Instituto Tecnológico de Buenos Aires (ITBA), Buenos Aires 1441, Argentina; jvillar@itba.edu.ar (A.J.V.); jsantos@itba.edu.ar (J.M.S.)
* Correspondence: rramele@itba.edu.ar
† Current address: C1437FBH Lavarden 315, Ciudad Autónoma de Buenos Aires 1441, Argentina.

Received: 27 September 2018; Accepted: 13 November 2018; Published: 16 November 2018

Abstract: The Electroencephalography (EEG) is not just a mere clinical tool anymore. It has become the de-facto mobile, portable, non-invasive brain imaging sensor to harness brain information in real time. It is now being used to translate or decode brain signals, to diagnose diseases or to implement Brain Computer Interface (BCI) devices. The automatic decoding is mainly implemented by using quantitative algorithms to detect the cloaked information buried in the signal. However, clinical EEG is based intensively on waveforms and the structure of signal plots. Hence, the purpose of this work is to establish a bridge to fill this gap by reviewing and describing the procedures that have been used to detect patterns in the electroencephalographic waveforms, benchmarking them on a controlled pseudo-real dataset of a P300-Based BCI Speller and verifying their performance on a public dataset of a BCI Competition.

Keywords: electroencephalography; brain-computer interfaces; waveform; p300; SIFT; PE; MP; SHCC

1. Introduction

Current society is demanding technology to provide the means to realize the utopia of social inclusion for people with disabilities [1]. Additionally, as societies are aging [2] the incidence of neuromuscular atrophies, strokes and other invalidating diseases is increasing. Concurrently, the digital revolution and the pervasiveness of digital gadgets have modified the way people interact with the environment through these devices [3]. All this human computer interaction is based on muscular movement [4], but these trends are pushing this boundary beyond the confines of the body and beyond the limitation of human motion. A new form of human machine communication which directly connects the Central Nervous System (CNS) to a machine or computer device is currently being developed: Brain Machine Interfaces (BMI), Brain Computer Interfaces (BCI) or Brain-Neural Computer Interfaces (BNCI).

At the center of all this hype, we can find a hundredth year old technology, rock-solid as a diagnosis tool, which greatly benefited from the shrinkage of sensors, the increase in computer power and the widespread development of wireless protocols and advanced electronics: the Electroencephalogram (EEG) [5].

EEG sensors are wearable [6] non-invasive, portable and mobile [7], with excellent temporal resolution, and acceptable spatial resolution [8]. This humble diagnosis device is been transformed into currently the best approach to detect, out-of-the lab in an ambulatory context, information from the Central Nervous System and to use that information to volitionally drive cars, steer drones, write emails, control wheelchairs or to assess alcohol consumption [9–12].

The clinical and historical tactic to analyze EEG signals is based on detecting visual patterns out of the EEG trace or polygraph [8]: multichannel signals are extracted and continuously plotted over a piece of paper. Electroencephalographers or Electroencephalography technician decode and detect patterns along the signals by visually inspecting them [5]. Nowadays clinical EEG still remains a visually interpreted test [8].

The need of quantitative procedures to automate the decoding of EEG signals has been materialized in BCI where around 71.2% is based on noninvasive EEG [4]. However, methods of decoding signals based on the detection of waveforms has been scarce. Hence, the traditional and knowledgeable approach has been neglected particularly in BCI Research. We aim to help fix this gap by providing a review of the methods which emphasize the waveform, the shape of the EEG signal and which can decode them in a supervised and semi-automated procedure.

The aim of this study is threefold: first to review current literature of EEG processing techniques which are based on analysis of the waveform. The second is to evaluate and study these methods by analyzing its classification performance against a pseudo-real dataset. And third, to verify their applicability to a real and public dataset.

This article unfolds as follows: Section 2 provides a brief introduction to EEG and the particularities of the EEG waveform characterization. Section 3.1 explains the waveform-based algorithms that are analyzed. In Section 3.6 the experimentation procedure is explained. Results are presented in Section 4 and finally Discussion and Conclusions are expounded in the final sections.

2. Electroencephalography

The Electroencephalography consists on the measurement of small variations of electrical voltage over the scalp. It is one of the most widespread used methods to capture brain signals and was initially developed by Hans Berger in 1924 and has been extensively used for decades to diagnose neural diseases and other medical conditions.

The first characterization that Dr. Berger detected was the Visual Cortical Alpha Wave, the *Berger Rythm* [13]. He understood that the amplitude and shape of this rhythm was coherently associated to a cognitive action (eyes closing). We should ask ourselves if the research advancement that came after that discovery would have happened if it weren't so evident that the shape alteration was due to a very simple and verifiable cognitive process.

The EEG signal is a highly complex multi-channel time-series. It can be modeled as a linear stochastic process with great similarities to noise [14]. It is measured in microvolts, and those slightly variations are contaminated with heavy endogenous artifacts and exogenous spurious signals. Figure 1 shows 5 s of a sample 8-channel EEG signal.

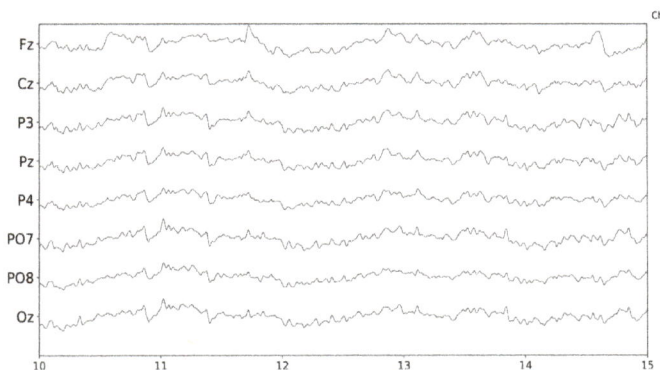

Figure 1. Sample EEG signal obtained from g.Tec g.Nautilus. Time axis is in seconds and five seconds are displayed. The eight channels provided by this device are shown.

The device that captures these small variations in potential differences over the scalp is called the Electroencephalograph. Electrodes are located in predetermined positions over the head, usually embedded in saline solutions to facilitate the electrophysiological interface and are connected to a differential amplifier with a high gain which allows the measurement of tiny signals. Although initially analog devices were developed and used, nowadays digital versions connected directly to a computer are pervasive. A detailed explanation on the particularities and modeling of EEG can be obtained from [15], and a description of its electrophysiological aspects from [16].

Overall, EEG signals can be described by their phase, amplitude, frequency and *waveform*. The following elements regularly characterize EEG signals:

- Artifacts: These are signal sources which are not generated from the CNS, but can be detected from the EEG signal. They are called endogeneous or physiological when they are generated from a biological source like face muscles, ocular movements, etc., and exogeneous or non-physiological when they have an external electromagnetic source like line induced currents or electromagnetic noise [17].
- Non-Stationarity: the statistical parameters that describe the EEG as a random process are not conserved through time, i.e., its mean and variance, and any other higher-order moments are not time-invariant [13].
- DC drift and trending: in EEG jargon, which is derived from concepts of electrical amplifiers theory, Direct Current (DC) refers to very low frequency components of the EEG signal which varies around a common center, usually the zero value. DC drift means that this center value drifts in time. Although sometimes considered as a nuisance that needs to get rid of, it is known that very important cognitive phenomena like Slow Cortical Potentials or Slow Activity Transients in infants do affect the drift and can be used to understand some particular brain functioning [5].
- Basal EEG activity: the EEG is the compound summation of myriads of electrical sources from the CNS. These sources generate a baseline EEG which shows continuous activity with a small or null relation with any concurrent cognitive activity or task.
- Inter-subject and intra-subject variability: EEG can be affected by the person's behavior like sleep hygiene, caffeine intake, smoking habit or alcohol intake previously to the signal measuring procedure [18].

Regarding how the EEG activity can be related to an external stimulus that is affecting the subject, it can be considered as

- Spontaneous: generally treated as noise or basal EEG.
- Evoked: activity that can be detected synchronously after some specific amount of time after the onset of the stimulus. This is usually referred as time-locked. In contrast to the previous one, it is often called Induced activity.

Additionally, according to the existence of a repeated rhythm, the EEG activity can be understood as

- Rhythmic: EEG activity consisting in waves of approximately constant frequency. It is often abbreviated RA (regular rythmic activity). They are loosely classified by their frequencies, and their naming convention was derived from the original naming used by Hans Berger himself, and after Alpha Waves (10 Hz), it came Delta (4 Hz), Theta (4–7 Hz), Sigma (12–16 Hz), Beta (12–30 Hz) and Gamma (30–100 Hz).
- Arrhythmic: EEG activity in which no stable rhythms are present.
- Dysrhythmic: Rhythms and/or patterns of EEG activity that characteristically appear in patient groups and rarely seen in healthy subjects.

The number of electrodes and their positions over the scalp determines a **Spatial Structure**: signal elements can be generalized, focal or lateralized, depending on in which channel (i.e., electrode) they are found.

EEG Waveform Characterization

The shape of the signal, the waveform, can be defined as the graphed line that represents the signal's amplitude plotted against time. It can also be called EEG biomarker, EEG pattern, signal shape, signal form and a morphological signal [13].

The signal context is crucial for waveform characterization, both in a spatial and in a temporal domain [13]. Depending on the context, some specific waveform can be considered as noise while in other cases is precisely the element which has a cognitive functional implication.

A waveform can have a characteristic shape, a rising or falling phase, a pronounced plateau or it may be composed of ripples and wiggles. In order to describe them, they are characterized by its amplitude, the arch, whether they have (non)sinusoidal shape, by the presence of an oscillation or imitating a sawtooth (e.g., Motor Cortical Beta Oscillations). The characterization by their sharpness is also common, particularly in Epilepsy, and they can also be identified by their resemblance to spikes (e.g., Spike-wave discharge).

Depictions may include subjective definitions of sharper, arch comb or wicket shape, rectangular, containing a decay phase or voltage rise, peaks and troughs, short term voltage change around each extrema in the raw trace. Derived ratios and indexes can be used as well, like peak and trough sharpness ratio, symmetry between rise and decay phase and slope ratio (steepness of the rise period to that of the adjacent decay period). For instance, wording like "Central trough is sharper and more negative that the adjacent troughs" [19] are common in the literature.

Other regular characterizations which are based on the waveform shape may encompass:

- Attenuation: Also called suppression or depression. Reduction of amplitude of EEG activity resulting from decreased voltage. When activity is attenuated by stimulation, it is said to have been "blocked" or to show "blocking".
- Hypersynchrony: Seen as an increase in voltage and regularity of rhythmic activity, or within the alpha, beta, or theta range. The term suggest an increase in the number of neural elements contributing to the rhythm, or in the synchronization of different neurons with the same discharge pattern [20].
- Paroxysmal: Activity that emerges from background with a rapid onset, reaching frequently high voltage and ending with an abrupt return to lower voltage activity.
- Monomorphic: Activity appearing to be composed of one dominant waveform pattern.
- Polymorphic: Activity composed of multiple frequencies that combine to form a complex waveform.
- Transient/Component: An isolated wave or pattern that is distinctly different from background activity.

The traditional clinical approach to study electroencephalographic signals consists in analyzing the paper strip that is generated by the plot of the signal obtained from the device. Expert technician and physicians analyze visually the plots looking for specific patterns that may give a hint of the underlying cognitive process or pathology. Atlases and guidelines were created in order to help in the recognition of these complex patterns. Video-electroencephalography scalp recordings are routinely used as a diagnostic tools [21] . The clinical EEG research has also focused on temporal waveforms, and a whole branch of electrophenomenology has arisen around EEG *graphoelements* [5].

Sleep Research has been studied in this way by performing Polysomnographic recordings (PSG) [22,23]. The different sleep stages are evaluated by visually marking waveforms or graphoelements in long-running electroencephalographic recordings, looking for patterns based on standardized guidelines [24]. Visual characterization includes the identification or classification of certain waveform components based on a subjective characterization (e.g., positive or negative peak polarity) or the location within the strip. It is regular to establish an amplitude difference between different waveforms from which a relation between them is reckoned and a structured index is created (e.g., sleep K-Complex is well characterized based on rates between positive vs negative amplitude) [25]. Other relevant EEG patterns for sleep stage scoring are alpha, theta, and delta waves, sleep spindles, polysplindles, vertex sharp waves (VSW), and sawtooth waves (REM Sleep).

Moreover, EEG data acquisition is a key procedure during the assessment of patients with focal Epilepsy for potential seizure surgery, where the source of the seizure activity must be reliably identified. The onset of the Epileptic Seizure is defined as the first electrical change seen in the EEG rhythm which can be visually identified from the context and it is verified against any clinical sign indicating seizure onset. The Interictal Epileptiform Discharges (IEDs) are visually identified from the paper strip, and they are also named according to their shape: spike, spike and wave or sharp-wave discharges [26].

Waveform characterization is the method in which analysis has been performed for Event Related Potentials (ERP). These are transient signal elements that may arise as a brain response to an external visual, tactile or auditory stimulus. ERPs are regularly used to assess auditory response in infants. They are extensively used and studied in Cognitive Neuroscience [27]. ERPs are identified by their components which are recognizable signal shapes assigned to the observed waveform, that can be linked to some cognitive or measurable psychological process. One of the most studied ERP is the P300, discovered in 1965 by Sutton, Braren, Zubin and John. This component is a positive deflection of a subject's EEG signal that arises when an unexpected and infrequent stimulus appears [1]. The P300 is widely utilized in BCI because it can be harnessed to implement a Speller application. Hence, P300 ERPs are a target phenomena to study by automatic waveform recognition methods.

Table 1 summarizes a list of depictions used to describe waveforms in the surveyed literature. Epilepsy has been described by the nature of oscillatory characterization of their waves, like ripples and wiggles, imitating sawtooths or by their geometric shape. For ERPs on the other hand, more elaborate indexes has been provided, establishing relations between amplitudes of signal components. Finally, Sleep studies and ICU research are areas where the most complex indexes has been derived, particularly the coupling of signal properties like phase, amplitude and frequency.

Table 1. EEG waveforms descriptions found in the surveyed literature.

Method	Phenomena	Reference
Positive Rounded Component	α-Waves, Epilepsy	[5,28]
Rising and Falling Phase	Epilepsy	[14,28]
Terminal plateau	Epilepsy	[14]
Ripples and Wiggles	Epilepsy, ERP	[14,26,29,30]
Sinusoidal Shape	Epilepsy	[19,28–31]
Sawtooth	Motor Imagery, Sleep	[22,26,28]
Sharpness or Spike-like	Epilepsy	[8,14,26,32]
Rectangular	Epilepsy	[14,19]
Line length	Anomaly Detection	[33]
Root Mean Square	Anomaly Detection	[33]
Wicket Shape	Epilepsy	[5,8,19,26,28,32]
Peak and Trough Sharpness Ratio	Epilepsy	[8,19,32,34]
Symmetry between rise and decay phase	Epilepsy	[8,19]
Slope Ratio	Sleep	[35]
Positive/Negative Peak Amplitude	ERP	[8,14,19,28,36,37]
Positive vs Negative Ratio	Sleep K-Complex	[26]
Base-to-Peak Amplitude	ERP	[19]
Peak-to-Peak Amplitude	ERP	[33,36]
Positive/Negative Peak Latency	ERP	[36]
Integrated Activity	ERP, Epilepsy, ICU	[25,33,38]
Cross-Correlation	ERP, Epilepsy, Sleep	[29,38]
Coupling		
Cross Frequency, Phase-Amplitude, Phase-Phase	Sleep	[19]
Period Amplitude Analysis	ERP, Epilepsy	[25,29,38]

Brain Sci. **2018**, *8*, 199

3. Materials and Methods

The exploration of methods based on waveforms is conducted by following the PRISMA [39] guidelines. Search is performed on Google Scholar, Semantic Web and IEEE Xplore search engines by the terms "Waveforms" OR "Shape" OR "Morphology" OR "Visual inspection" + "EEG".

The following criteria is proposed to identify methods which are based on the signal's waveform:

1. The analysis considers the shape of the plot of the signal.
2. The pattern can be identified and verified by visual inspection.
3. The pattern matching is performed in time-domain.
4. The method encompass a feature extraction procedure.
5. The feature extraction procedure allows to create a template dictionary.

As described in [40] the Pattern Matching problem in Signal processing is finding a signal given the region that best describes the structure of the prototype signal template. On the other hand, a *feature* is a meaningful quantification, usually a multidimensional vector, that synthesize the information of a given signal or signal segment [41].

3.1. EEG Waveform Analysis Algorithms

Shape or waveform analysis methods are considered as nonparametric methods. They explore signal's time-domain metrics or even derive more complex indexes or features from it [42].

One of the earliest approach to automatically process EEG data is the Peak Picking method. Although of limited usability, peak picking has been used to determine latency of transient events in EEG [43,44]. Straightforward in its implementation, it consists in assigning a component to a simple waveform element based on the expected location of its more prominent deflection [31]. Of regular use in ERP Research, the name of many of the EEG features reference directly a peak within the component, e.g., P300 or P3a P3b or N100. This leads to a natural way to classify them visually by selecting appropriate peaks and matching their positions and amplitudes in an orderly manner. The letter provides the polarity (Positive or Negative) and the numbering shows the time referencing the stimulus onset, or the ordinal position of each peak (first, second, etc).

A related method is used in [45] where the area under the curve of the EEG is sumarized to derive a feature. This was even used in the seminal work of Farwell and Donchin on the P300 Speller [41,46]. Additionally, a logarithmic graph of the peak-to-peak amplitude which is called amplitude integrated EEG (aEEG) [38] is used nowadays in Neonatal Intensive Care Units.

Other works on EEG explored the idea to extend human capacities analyzing EEG waveforms. In [47] a feature derived from the amplitude and frequency of its signal and its derivative in time-domain is used. Moreover, Yamaguchi et al. [48] explored the use of Mathematical Morphology, where the time-domain structure of contractions and dilations were studied. Finally the proposals of Burch, Fujimori, Uchida and the Period Amplitude Analysis (PAA) [49] algorithm are few of the earliest depictions where the idea of capturing the shape of the signal were established.

According to the defined criteria, the algorithms that will be evaluated are as follows:

- Matching Pursuit
- Permutation Entropy
- Slope Horizontal Chain Code
- Scale Invariant Feature Transform

All these methods provide a feature f that can be used as a template. The notation $f = \{f_i\}_1^n$ or $f = \{f_i\}_{i \in J}$ is used to describe the concatenation of scalar values to form a multidimensional feature vector $f = \{f_1, f_2, ..., f_n\}$. These algorithms are all based on metrics that are extracted from the shape of the single channel digital EEG signal $x(n)$, with n varying from 1 to the length N of the EEG segment in sample points. These features are used to create dictionaries or template databases. Finally, these templates provide the basis for the pattern matching algorithm and offline classification.

Algorithms were implemented on MATLAB 2014a (Mathworks Inc., Natick, MA, USA). To maintain reproducibility, the dataset described in Section 3.6.1 and the source code has been made available in the online repository of the Code Ocean platform under the name *EEGWave*.

3.2. Matching Pursuit—MP 1 and MP 2

Pursuit algorithms refer, in their many variants, as blind source separation [50] techniques that assume the EEG signal as a linear combination of different sparse sources extracted from a template's dictionaries. Matching Pursuit *MP* [51], the most representative of these algorithms, is a greedy variant that decomposes a signal into a linear combination of waveforms, called atoms, that are well localized in time and frequency [52]. Given a signal, this optimization technique, tries to find the indexes of m atoms and their weights (contributions) that minimize,

$$\varepsilon = \left\| x(n) - \sum_{i=1}^{m} w_i g_i(n) \right\| \tag{1}$$

which is the error between the signal and its approximation constructed by the weighted w_i atoms g_i, and calculating the euclidean norm $\|\cdot\|_2$. The algorithm goes by first setting the approximating signal $\tilde{x}_0$ as the original signal itself,

$$\tilde{x}_0(n) = x(n) \tag{2}$$

and setting the iterative counter k as 1. Hence, it searches recurrently the best template out of the dictionary that matches current approximation.

$$g_k = \arg\max_{g_i} \left| \sum_{n=1}^{N} \tilde{x}_{k-1}(n) \, g_i(n) \right| \tag{3}$$

where g_i are all the available scaled, translated and modulated atoms from the dictionary. The operation $|\cdot|$ corresponds to the absolute value of the inner product. This step determines the atom selection process, and their contribution is calculated based on

$$w_k = \frac{\sum_{n=1}^{N} \tilde{x}_{k-1}(n) \, g_k(n)}{\|g_k\|^2} \tag{4}$$

with k representing the index of the selected atom g_k and $\|\cdot\|_2$ its euclidean norm. Finally the contribution of each atom is subtracted from the next approximation [32,51,53]

$$\tilde{x}_k(n) = \tilde{x}_{k-1}(n) - w_k g_k(n) \tag{5}$$

The stopping criteria can be established based on a limiting threshold on Equation (1) or based on a predetermined number of steps and selected atoms. Two variants of this algorithm are evaluated. In *MP 1* the dictionary is constructed with the normalized templates directly extracted from the real signal segments which is a straightforward implementation of the pattern matching technique. In *MP 2* the coefficients of Daubechies least-asymmetric wavelet with 2 vanishing moments atoms are used to construct the dictionary [54]. For the first version, the matching against the template is evaluated according to Equation (1) directly, whereas for the latter each feature is crafted by decomposing the signal in its coefficients and building, an eventually sparse, vector with them:

$$f = \left\{ w_i \right\}_1^D \tag{6}$$

where D is the size of the dictionary.

3.3. Permutation Entropy—PE

Bond and Pompe Permutation Entropy has been extensively used in EEG processing, with applications on Anesthesia, Sleep Stage evaluation and increasingly for Epilepsy pre-ictal detection [55]. This method generates a code based on the orderly arrangement of sequential samples, and then derives a metric which is based on the number of times each sequence is found along the signal. This numeric value can be calculated as information entropy [56]. Let's consider a signal on a window of length W represented by the sample points

$$(x_1, x_2, ..., x_W) \tag{7}$$

and resampled by τ intervals, starting from the sampling point n, doing

$$(x_n, x_{n+\tau}, x_{n+2\tau}..., x_{n+(m-1)\tau}). \tag{8}$$

This sequence is of order m, which is the number of sample points used to derive the ordinal element called π. There are $m!$ ways in which this sequence can be orderly arranged, according to the position that each sample point holds within the sequence in a decreasing order relationship [57]. For example if $m = 3$, and the first sample point is the bigger, the second is the smaller and the third one is in the middle, the ordinal element π corresponds to $(1, 3, 2)$. Thus, along the signal window there can be at most k different ordinal (and overlapping) elements π_s

$$(\pi_1, \pi_2, ..., \pi_k) \tag{9}$$

with $k = W - (m-1)\tau$. The probability density function *pdf* for all the available permutations of order m should be $\mathbf{p} = (p_1, p_2, ..., p_{m!})$ with $\sum_{i=1}^{m!} p_i = 1$.

Hence, the time series window is mapped to a new set of k ordinal elements, and the *pdf* can be calculated by the empirical permutation entropy,

$$p_i = \frac{1}{k} \sum_{s=1}^{k} [\pi_s = \pi_i] \tag{10}$$

with $1 \leq i \leq m!$. The Iverson Bracket $[\cdot]$ resolves to 1 when their logical proposition argument is true, 0 otherwise. Therefore, for each i only those ordinal elements π_s that were effectively found along the signal are counted to estimate p_i, and zero elsewhere. The empirical permutation entropy can be calculated from the histogram as,

$$H(\mathbf{p}) = \sum_{i=1}^{m!} p_i \, log \frac{1}{p_i}. \tag{11}$$

The implemented code was derived from [58], and the model description from [59]. This procedure produces a scalar number for the given signal window of size W. To derive a feature, a sliding window procedure must be implemented to cover an entire segment of length N. Thus, the length of the feature is $N - (W + \tau(m-1))$.

$$f = \left\{ H(\mathbf{p})_u \right\}_{W+\tau m}^{N}. \tag{12}$$

with u varying on a sample by sample basis along the signal, starting from the specified index.

3.4. Slope Horizontal Chain Code—SHCC

This algorithm [45] proceeds by generating a coding scheme from a sequence of sample points. This encoding is based on the angle between the horizontal line on a 2D-plane and any segment produced by two consecutive sample points, regarding them as coordinates on that plane.

A signal of length N, can be represented by a list of ordered-pairs e,

$$e = [(x,y)_1, (x,y)_2, ..., (x,y)_N] \tag{13}$$

and it can be divided into G different blocks. These blocks are obtained by resampling the original signal from the index

$$G = \lfloor n + (m\Delta) + 0.5 \rfloor \tag{14}$$

with n being the original sampling index on $1 \leq n \leq N$ and $\lfloor \cdot \rfloor$ being the floor operation, i.e., rounding of the number argument to the closest smaller integer number. On the other hand, Δ can be obtained by

$$\Delta = \left\lceil \frac{N}{G+1} \right\rceil \tag{15}$$

with $G < N$ and using instead $\lceil \cdot \rceil$ as the ceil operation, the rounding to the closest bigger integer number. Lastly, the value m can be derived from

$$m = sign\left(\frac{N-1}{\Delta}\right)\left\lfloor \left|\frac{N-1}{\Delta}\right| \right\rfloor. \tag{16}$$

This resampling produces a new sequence of values,

$$e' = [(x',y')_1, ..., (x',y')_s, ..., (x',y')_G]. \tag{17}$$

The next step is the normalization of each ordered-pair as vectors $\mathbf{x}' = (x'_1, ..., x'_G)$ and $\mathbf{y}' = (y'_1, ..., y'_G)$ according to

$$\hat{\mathbf{x}} = \frac{\mathbf{x}' - \min(\mathbf{x}')\mathbf{1}}{\max(\mathbf{x}') - \min(\mathbf{x}')} \tag{18}$$

$$\hat{\mathbf{y}} = \frac{\mathbf{y}' - \min(\mathbf{y}')\mathbf{1}}{\max(\mathbf{y}') - \min(\mathbf{y}')} \tag{19}$$

with $\mathbf{1}$ being the vector of length G with all their components equal to 1. Hence, the scalar components $\hat{x}_s$ of $\hat{\mathbf{x}}$, and $\hat{y}_s$ of $\hat{\mathbf{y}}$, with s varying between 1 and G are effectively normalized to $\hat{x}_s, \hat{y}_s \in [0,1]$.

Finally, the feature is constructed by calculating the point-to-point slope against the horizontal plane,

$$f = \left\{ \frac{\hat{y}_s - \hat{y}_{s-1}}{\hat{x}_s - \hat{x}_{s-1}} \right\}_2^G \tag{20}$$

3.5. Scale Invariant Feature Transform—SIFT

SIFT [60] is a very successful feature extraction technique from Computer Vision. It has a biomimetic inspiration on how the visual cortex analyze images based on orientations [61]. This method has been used in [62] to analyze EEG signals based on their plots on 2D images.

The first step of the algorithm is the plot generation based on single-channel EEG segments $x(n)$. Hence, this signal is normalized by the z-score [63]:

$$\tilde{x}(n) = \left\lfloor \frac{\delta(x(n) - \bar{x})}{\sigma_x} \right\rceil \tag{21}$$

with δ being the signal magnification factor and $\bar{x}$ and σ_x, the mean and standard deviation of x on the signal segment. The width of the image is determined based on the 1-s length size of the segment in sample units. This corresponds to the sampling frequency F_s of the EEG signal segment. The width is adjusted by multiplying by the magnification factor δ,

$$w = \delta \, F_s \tag{22}$$

whereas the height is calculated based on the peak-to-peak amplitude of the signal within the segment,

$$h = \max_n \tilde{x}(n) - \min_n \tilde{x}(n). \tag{23}$$

Equation (24) determines the vertical position of the image where the signal's zero value will be located.

$$z = \left\lfloor \frac{h}{2} \right\rfloor - \left\lfloor \frac{\max_n \tilde{x}(n) + \min_n \tilde{x}(n)}{2} \right\rfloor. \tag{24}$$

Finally, a binary, black-and-white image plot is generated based on

$$I(z_1, z_2) = \begin{cases} 255 & \text{if } z_1 = \delta \, n; \; z_2 = \tilde{x}(n) + z \\ 0 & \text{otherwise} \end{cases} \tag{25}$$

where z_1 and z_2 are the image coordinates values, 255 represents white and 0 is the background black color of the plot. These points are interpolated using the Bresenham algorithm [62].

Once the plot is generated, its center is used to localize the center of the SIFT patch. This region of the image, where the signal's most important salient waveform should be located, is divided in a grid of 4×4 block and the bidimensional gradient vectors are calculated on each one of them. Therefore, for each block (i, j) within the patch, a histogram $h(i, j, \theta)$ of the gradient orientations, for 8 circular orientations θ, are calculated. This histogram is concatenated for all the 16 blocks and a feature is thus formed:

$$f = \left\{ \left\{ \left\{ h(i, j, \theta) \right\}_{i \in I} \right\}_{j \in I} \right\}_{\theta \in \Theta} \tag{26}$$

with i and j belonging to $I = \{0, 1, 2, 3\}$ and localizing the 16 blocks within the grid. The angles θ that belong to Θ are the eight possible equidistant values between 0 and 315. This vector is normalized, clamped to 0.2, and re-normalized again. Details of the method can be found on [60,62]. It was implemented using the VLFeat [64] public Computer Vision libraries.

3.6. Experimental Protocol

The objective of the following experiments is to assess the performance of the algorithms that aim to recognize the shape of the P300 waveform, obtained after averaging signal segments. This performance is evaluated by processing a pseudo-real dataset with two modalities where subtle alterations on the latency and amplitude of the P300 component are simulated in a controlled environment. The experiments are performed by the offline evaluation of the character identification rate of a Visual P300-Based BCI Speller application.

Farwell and Donchin P300 Speller [46,65] is one the most used BCI paradigms to implement a thought translation device and to send commands to a computer in the form of selected letters, similar to typing on a virtual keyboard. This procedure exploits a cognitive phenomena raised by the oddball paradigm [27]: along the EEG trace of a person which is focusing on a sequence of two different visual flashing stimulus, a particular and distinctive transient component is found each time the expected stimulus flashes. This is cleverly utilized in the P300 Speller, where rows and columns of a 6×6 matrix flashes randomly but only the flashing of a column or row where the letter that a user is focusing will trigger concurrently the P300 ERP along the EEG trace.

A problem with the information produced by a P300 Speller is that the subjects that take part on the experiment are within the closed loop of the BCI system and the human is not a static compliant entity that always performs what the experimenter asks for in a precise and consistent way [66].

Therefore, P300 experiments data is often mined with *null-signals*. These are EEG streams which are marked as having the signal component but, because the subject was not particularly focused, or concentrated, the expected signal element is not generated. This lack of certainty may be in detriment of any conducted analysis and can be misleading or difficult to deal with. Previous works have addressed this same issue, particularly when benchmarking different algorithms [31,43,67].

In order to tackle this problem, a pseudo-real dataset based on an EEG stream is generated under two different modalities. A passive modality and an active modality.

3.6.1. EEG Stream Generation

Eight (8) healthy participants are recruited voluntarily and the experiment is conducted anonymously in accordance with the Declaration of Helsinki published by the World Health Organization. No monetary compensation is handed out and they agree and sign a written informed consent. This study is approved by the *Departamento de Investigación y Doctorado, Instituto Tecnológico de Buenos Aires (ITBA)*. The participants are healthy and have normal or corrected-to-normal vision and no history of neurological disorders. These voluntary subjects are aged between 20–40 years old. EEG data is collected in a single recording session. Each subject is seated in a comfortable chair, with her/his vision aligned to a computer screen located one meter in front of her/him. The handling and processing of the data and stimuli is conducted by the OpenVibe platform [68]. Gel-based active electrodes (g.LADYbird, g.Tec, Austria) are used on locations Fz, Cz, Pz, Oz, P3,P4, PO7 and PO8 according to the 10–20 international system. Reference is set to the right ear lobe and ground is preset as the AFz position. Sampling frequency is set to 250 Hz.

The experimental protocol is composed of 35 trials to spell 7 words of 5 letters each. Each trial is composed of 10 intensification sequences of the 6 columns and 6 rows of the Speller Matrix. This yields exactly 120 intensifications of rows and columns per trial. The duration of each intensification as well as the Inter-Stimulus Interval, the pause between stimulations, are set to 0.125 s. This provides a 4 Hz frequency of flashes on the screen. The initial pause and the inter-trial pauses are set to 20 s. The whole experiment lasts for around 1400 s. This produces an EEG stream which contains 4200 marked sections where 3500 of them are labeled as *True* and the remaining 700 as *False*. The extracted EEG signals are band-pass filtered using a 4th order Butterworth digital filter between 0.1 and 10 Hz and a 50 Hz notch filter is applied to remove line AC noise. The EEG trace is finally downsampled to 16 Hz. Segments of 1-s length are extracted according to the markers information and those with variations larger than 70 μV are identified as artifacts and eliminated.

Four out of the eight participants are instructed to passively watch the flashing screen while not focusing on any particular letter. They do not receive any extra information on the screen. None of them have any experience with a BCI device. A questionnaire is handed out at the end of the experiment with questions about how the participant felt during it, without giving more details.

The remaining four participants perform a copy-spelling task where the computer monitor highlights the target letter, which is the one that the subject needs to focus. Across the duration of the trial, the current target letter is informed at the bottom of the screen.

3.6.2. Passive Modality

First for a passive modality, real P300 ERP templates obtained from a public dataset, are superimposed into the generated EEG stream of 4 subjects. A set of template ERPs is acquired from the Subject Number 8 of the public dataset 008-2014 [69] published on the BNCI-Horizon website [70] by IRCCS Fondazione Santa Lucia. The experimental protocol implemented to produce this dataset is the same as the one described in Section 3.6.1. On the other hand, the EEG traces where these templates are superimposed, are experimentally obtained by subjects which are observing the flashing of the stimulus matrix during a P300 Speller procedure but they do not engage in focusing on any letter in particular. Everything is there, except the P300 ERP component. Hence, along the EEG stream, the markers information is used to localize the *True* segments where the P300 should be found, and those timing locations are used to superimpose the extracted ERP waveform. By implementing this pseudo-real approach, it is possible to effectively control null-signals and to adjust the shape of the evoked potential.

A sample P300 ERP obtained from the trial number 2 of Subject 8 can be seen in Figure 2. These templates are selected due to their shapes more closely resembling the prototypical P300 waveform [71,72]. They are produced by extracting segments for this subject and by point-to-point coherently average them.

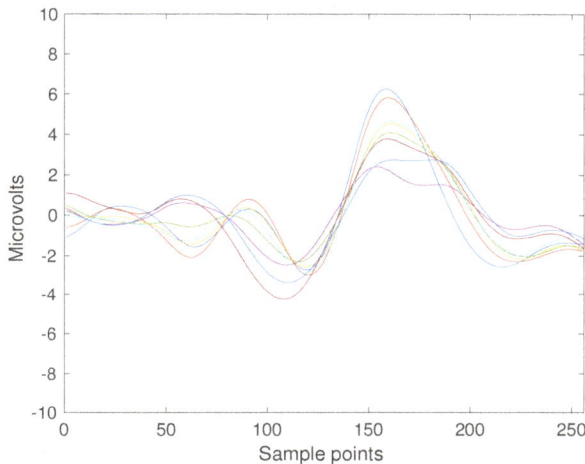

Figure 2. ERP Template obtained from the coherent point-to-point ensemble average from the signals of Subject Number Eight of the BNCI Horizon public dataset 008-2014. The template is 1 s long which is 256 sample points, and the eight channels are superimposed with different colors. The P3b component can be seen around the sample index 150 and 200.

3.6.3. Active Modality

Second, an active modality is also implemented, where a P300-Based BCI Speller experiment is performed on four subjects. For this scenario, the signal segments are modified to guarantee the inclusion of a P300 component. However, in this case the templates are extracted from the same subject. Hence, the EEG signal is preprocessed and labeled segments are extracted. Segments labeled *True* are coherently point-to-point averaged, and 70 templates are produced from the whole set of 35 trials.

Once templates are procured, a random *False* segment for the same subject is obtained. This is used as a baseline signal and is added to the template, conforming a new segment which has a superimposed P300 template. This procedure continues until the 700 segments marked as *True* are completed.

Figure 3 shows a 5 s sample of the EEG trace obtained with the MNE library [73]. Channel *S* represents the twelve different stimulus markers (columns or rows) while channel *L* represent the label (*True* vs *False*). Labels are represented by square signals. *False* segments are marked with single amplitude square signals while *True* segments are identified by double-amplitude square signals. Subfigure (a) shows the signals before the ERP template is superimposed while subfigure (b) shows the same signals with the superimposed ERP template. At first-sight, differences are really hard to spot visually. Subfigures (c) and (d) show only one second of channels Cz and L from the same segment. The superimposed ERP can be devised enclosed by the vertical bars, around 31.5 s, where in (d) the peak is slightly bigger. Figure 4 shows the obtained ensemble average ERPs as result of superimposing the template signal into the EEG stream, time-locked to the stimulus onset. These 12 point-to-point averaged segments correspond to the first trial of the EEG stream.

(**a**) EEG trace of the original signal. The horizontal axis represents 10 s of the EEG stream, from the 28th second up to the 38th.

(**b**) The same 10-s eight-channel signal segment with the superimposed template.

(**c**) EEG sample of Cz and L channel of the original EEG trace. Only 1 second is shown here, at the 31th second.

(**d**) The same segment with the superimposed template.

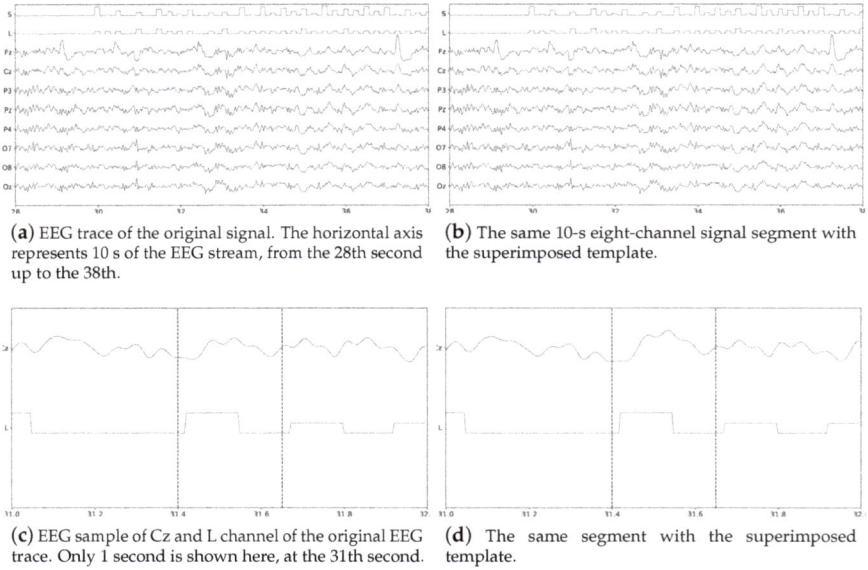

Figure 3. Eight-channel EEG signal for Subject Number 1 of the pseudo-real dataset without and with the superimposed ERP Template. The channel L, the mark which identifies where to superimpose the P300 ERP, is shown as well as the channel S which identifies the stimulus that was presented. On (**c**,**d**) the small variation that was introduced by the superimposition of the ERP can be seen enclosed by the vertical bars, where the slope of the bump on subfigure (**d**) is slightly bigger.

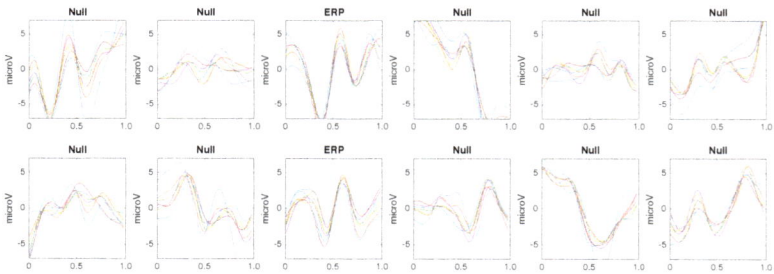

Figure 4. Point-to-point averaged signals. These are extracted from the first letter identification trial of the Subject 1 of the pseudo-real dataset. The ERP is superimposed on classes 3 and 9. Class 3 is obtained while averaging the segments where the row of the speller matrix is intensified whereas class 9 is calculated from the intensification of the corresponding column.

3.6.4. Experiments

The experiments are as follows:

- Experiment 1—Letter Identification Performance: the letter identification performance of each one of these methods on the artificially generated pseudo-real dataset. The pool of 70 P300 ERP waveforms, either obtained from the same subject in the passive-modality or from each subject in the active-modality are used to compose the artificial P300 wave in the pseudo-real dataset. Templates are randomly selected.

- Experiment 2—Latency Noise: Instead of superimposing the P300 ERPs over the EEG trace at the exact locations where stimulus onsets are situated, an artificial latency lag is added. The lagging value is picked from a uniform distribution $U(0, 0.4)$ [s] ranging from 0 to 0.4 of the 1 s segment size [74].
- Experiment 3—Component Amplitude Noise: the amplitude of the main P3b component of the ERP template is randomly altered. This component is defined to be located from the stimulus onset between 148 ms up to 996 ms which is around 840 ms long. This waveform element, multiplied by a gain factor, is subtracted from the original template. This gain factor between 0 and 1 is drawn from a uniform distribution $U(0, 1)$. Additionally this subtracted waveform is multiplied by a Gaussian window with a support of the same length [75]. This avoids adding any discontinuity into the artificial generated signal.

All these experiments are executed using cross validation procedure dividing the letter to spell in two sets, preserving the structure of the letter identification trials. Spelling letters are scrambled while the order and group of each intensification sequence is preserved.

Finally the performance at letter identifications for these same methods is evaluated by running an offline BCI Simulation on the Dataset IIb of the BCI Competition II (2003) [76]. The protocol of this dataset is very similar to what was used to obtain the pseudo-real dataset. The sampling frequency of this dataset is 240, the number of letters are 73 where the first 42 are used to create the template dictionary for all the methods and the remaining 31 are used to test the character recognition rate performance. Additionally, in this dataset the number of available intensification number sequences is 15. The classification method Support Vector Machine SVM with a linear kernel, is added for comparison as control using a feature f constructed by normalizing the signal on each channel [77]. This method has been proved efficient in decoding P300 in several BCI Competitions [78].

3.6.5. Classification

The same classification algorithm based on k-nearest neighbors is used for all the methods [79]. The experimental protocol used to generated the pseudo-real dataset used in the experiments 1 to 3 is composed of 35 trials to spell 7 words of 5 letters each. Each trial is composed of 10 intensification sequences of the 6 columns and 6 rows of the Speller Matrix. Fifteen trials are used to build the dictionary of templates, extracting the averaged EEG segments for the row and column that already contain the P300 ERP, hence shielding 30 different templates per channel. Figure 5 shows the set of templates while using the first 15 trials of the dataset.

Described algorithms produce a feature f for each averaged EEG segment. The aim of the classification procedure is to identify for the remaining 20 trials which of the 6 features f that were obtained for row intensification, labeled by $\{1, ..., 6\}$, and which of the 6 features for column intensification, named $\{7, ..., 12\}$ are the ones that elicited the P300 response on the averaged EEG segment. The row number of the matrix can be obtained by doing

$$r\hat{o}w = \arg \min_{u \in \{1, ..., 6\}} \sum_{i=1}^{k} \|f_u - q_i\|^2 \tag{27}$$

with q_i being the set of k-nearest neighbors of the feature f_u with u varying from 1 to 6. The parameter k represents the number of neighbors chosen from the dictionary of templates. The column can be obtained in the same way,

$$c\hat{o}l = \arg \min_{u \in \{7, ..., 12\}} \sum_{i=1}^{k} \|f_u - q_i\|^2 . \tag{28}$$

Thus, the letter identification performance can be obtained by measuring the accuracy channel-by-channel at identifying the correct letter on the matrix, coordinated by $r\hat{o}w$ and $c\hat{o}l$.

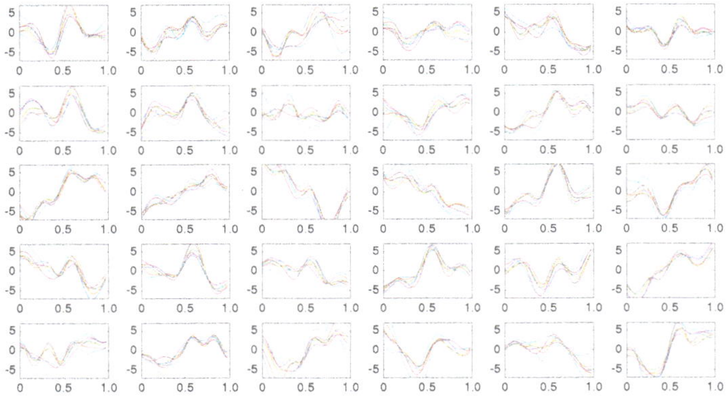

Figure 5. Coherently averaged signals segments of 1-second length containing the superimposed ERP. Vertical axis unit is µV. Each one is extracted from the EEG signal of the Subject 1 of the pseudo-real dataset. These averaged signals correspond to the 15 first trials (2 averaged signals from each trial, one belonging to the column flashing and the other to row flashing). These are the templates used to build a dictionary per channel per subject and are used by the classification algorithm described in Section 3.6.5.

4. Results

Results for the first experiment are shown in Figures 6 and 7. The performance while identifying each letter of the standard P300 Speller Matrix, and the channels where the best and worst performance are attained, are shown. Each one represents the percentage of letters that is actually predicted by the algorithms using a cross-validation procedure. As previously described the data is continuously divided in two sets, where the first 15 letters are used to derive the dictionary of templates while the remaining 20 letters are used to measure the letter identification performance. This is repeated one hundred times, and performances averaged. Figure 6 shows the results for the passive modality while Figure 7 shows the results for the active modality.

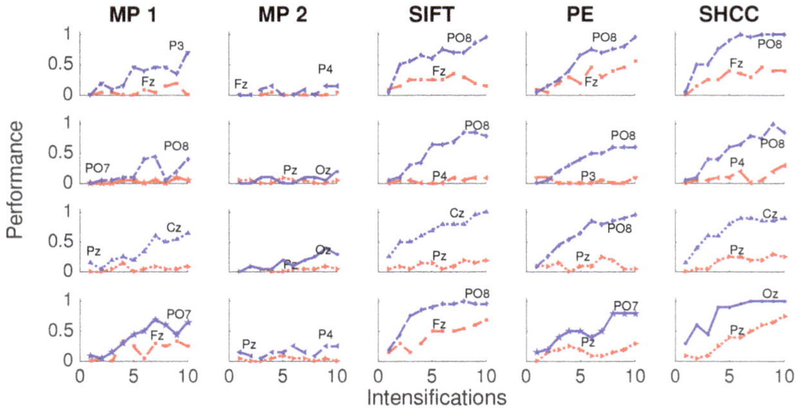

Figure 6. Passive Modality—Experiment 1: Speller performance curves obtained for each method for the four subjects that performed the passive modality protocol. Y-axis shows performance accuracy while X-axis shows the number of intensification sequences used to calculate the point-to-point signal average. The two curves show the performance for the best and worst performing channel.

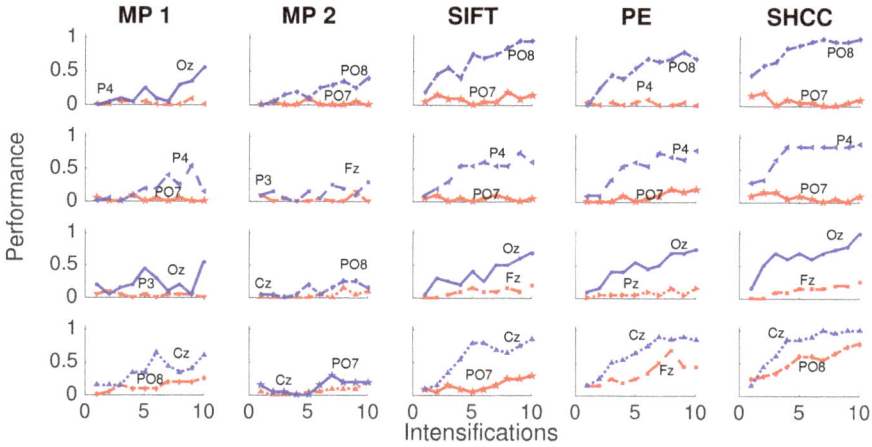

Figure 7. Active Modality—Experiment 1:Speller performance curves obtained for each method for the four subjects that performed the active modality protocol. Y-axis shows performance accuracy while X-axis shows the number of intensification sequences used to calculate the point-to-point signal average. The two curves show the performance for the best and worst performing channel.

Figures 8 and 9 shows the performance curves for five algorithms for the second experiment, where a noisy latency lag is included. Best and worst channels are also shown.

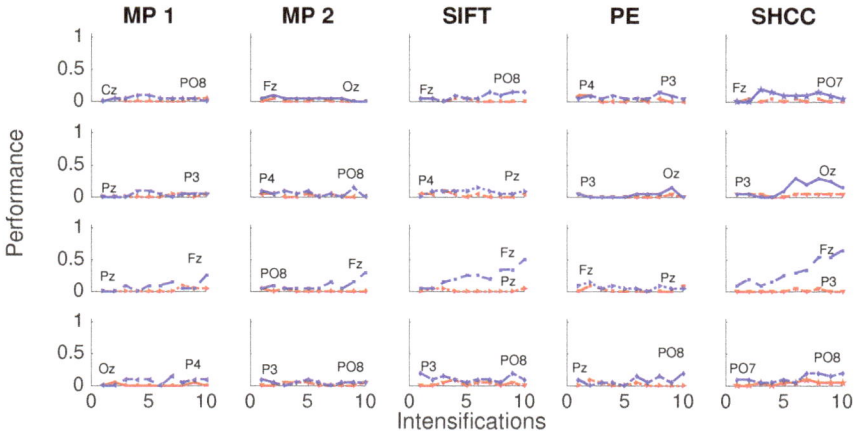

Figure 8. Passive Modality—Experiment 2: Performance curves for four subjects for the five algorithms when a random latency is included when superimposing the P300 signal template.

Finally, Figures 10 and 11 represents the performance values obtained for the Experiment 3, when the amplitude of the P3b component of the template is randomly attenuated.

Furthermore, results obtained for the dataset BCI Competition 2003 IIb are shown in Figure 12 and in Table 2. For this experiment the number of available intensification sequences is 15.

Table 2. Speller classification performance obtained for the dataset IIb of the BCI Competition II (2003) for each one of the algorithms using 15 repetitions of intensification sequences. The first 42 trials are used for training to build the template dictionary and the remaining 31 for testing. The channel where the best performance is attained, is also shown.

Method	Channel	Performance
MP 1	Cz	50%
MP 2	FC1	22%
SIFT	FC1	67%
PE	CP1	22%
SHCC	Cz	61%
SVM	C1	32%

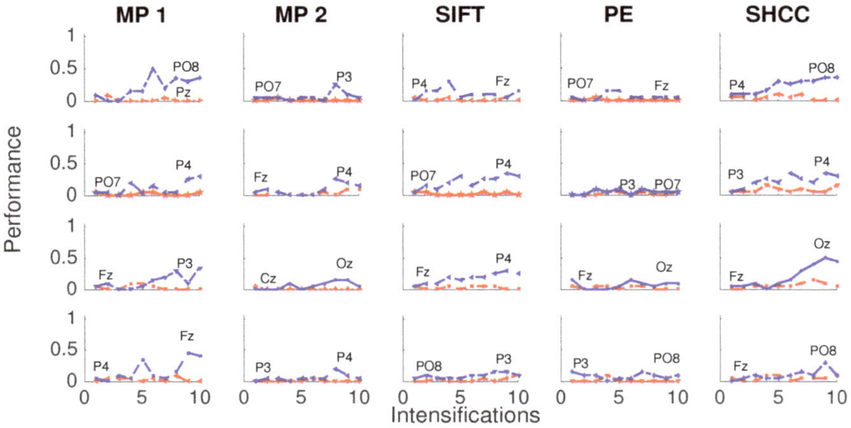

Figure 9. Active Modality—Experiment 2: Performance curves for the four subjects for the five algorithms. A random latency is included while superimposing the P300 signal template.

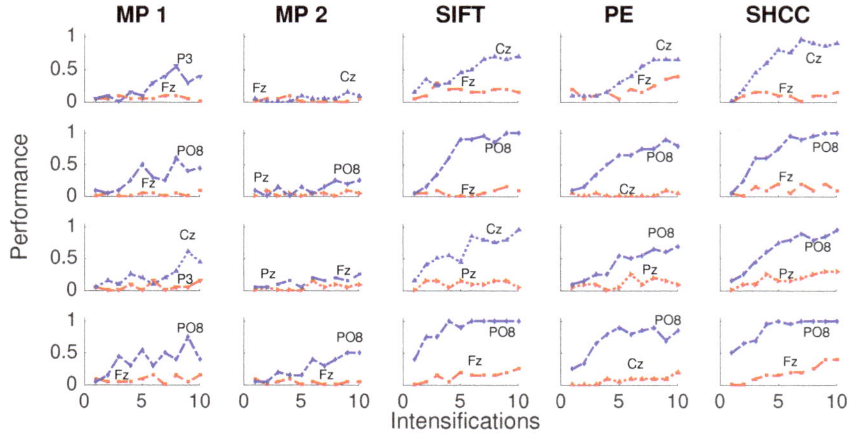

Figure 10. Passive Modality—Experiment 3:Performance curves for four subjects for the five algorithms when the amplitude of the P3b component of the template is randomly attenuated.

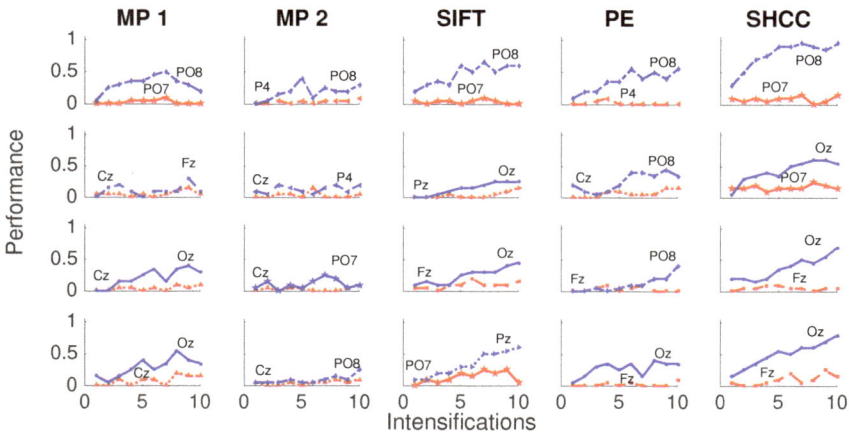

Figure 11. Active Modality—Experiment 3:Performance curves for four subjects for the five algorithms when the amplitude of the P3b component of the template is randomly attenuated.

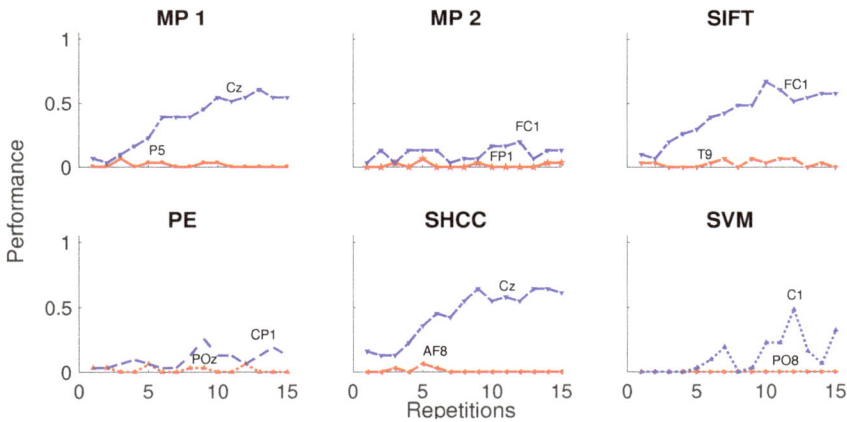

Figure 12. Speller performance obtained for the Dataset IIb of the BCI Competition II (2003) for each one of the algorithms. An offline BCI Simulation is performed using the first 42 trials as training and the remaining 31 as testing. The horizontal axis show the number of intensification sequences, from 0 to 15 for this dataset, while the vertical axis show the performance rate.

5. Discussion

A significant reduction of performance was found when latency noise is added. The latency noise reduces the information contained in the averaged signal, mainly due to the invalidation of the SNR enhancement performed by the signal averaging procedure. This reduction alters the obtained shape of the waveform of the ERP and impacts on the performances regardless of the method. On the other hand, all the algorithms show some resistance to noise in peak amplitudes of the main component. This is shown by the similarities of obtained results between the Experiment 1 and 3.

Using a straightforward dictionary of templates for *MP-1* proved more beneficial in terms of performance than the approach of using a Hilbert base of Wavelets atoms on *MP-2*. Either applying latency noise or amplitude noise, the method based on the signal's templates instead of using their coefficients achieve much better character identification rates.

Regarding results produced for the public and real dataset IIb of P300 ERP from the Berlin BCI Competition II (2003), the obtained character identification rate is above theoretical chance level, and for some algorithms close to the usable threshold of 70% [80,81]. When the character identification rate reaches this level of performance, the usage of word predicting algorithms allows to implement practical speller applications. Results for this competition have shown perfect classification with tailored algorithms [82]. This level is also similar to the performance obtained for the Experiment 3, which represents coincidentally the more realistic scenario for the pseudo-real dataset. It is important to remark that the algorithms presented here analyze the waveform structure of a single-channel signal [65,83].

6. Conclusions

The purpose of this work is threefold, (1) raise awareness about the utility of using automatic waveform-based methods to study EEG signals, (2) to provide an overview of the state-of-the-art of those methods, and (3) to compare those methods and verify if it is possible to obtain acceptable classification performances based exclusively on the signal's waveform.

The higher performance results are obtained for the methods *SHCC* and *SIFT* either on the pseudo-real dataset and on the BCI Competition. We verified that it is possible to obtain discriminating information from the underlying signal based exclusively on an automated method of processing the waveforms. This brings the possibility to use these techniques to implement intelligible [84] automatic detection procedures, i.e., systems that are able to emphasize clearly and noticeable what are the factors that caused the system action, decision or classification. This is due to the fact that they are based on metrics which can be visually verified.

Further work should be conducted in terms of a multichannel meaningful extension of these waveform-based methods [83]. Moreover, the possibilities of finding overcomplete dictionaries for matching pursuit sparse representation based on obtained signal templates, could also be considered an area of future improvement.

We believe that the adoption of a *hybrid* methodology which can process the signal automatically, but at the same time, maintains an inherent intelligible property that can be mapped to existing procedures, and above all, can maintain the clinician trust on the system behavior is beneficial to Clinical Practice, Neuroscience and BCI research. Additionally, this may foster collaboration in a multidisciplinary environment and may ease the acceptance and translation of BCI technology [66]. The reason being, for caregivers and medical staff, particularly those with the expertise of the clinical EEG which is based on waveforms, they may feel a natural understanding of how the system is performing.

Another benefit of these methodologies is that they have a potential universal applicability. As they are only analyzing waveforms, they can be explored in other disciplines where the structure or shape of the waveform is of relevance. Analyzing signals by their waveforms is relative common in chemical analysis [85], seismic analysis in Geology [86], and quantitative financial analysis. Electrocardiogram EKG, on the other hand, has been extensively processed and studied analyzing the waveform structure [87].

Author Contributions: This projects is part of a R.R.'s PhD Thesis which is directed by J.M.S. and codirected by A.J.V.

Funding: This project was supported by the ITBACyT-15 funding program issued by ITBA University.

Acknowledgments: We would like to thank to Valentina Unakafova for providing the Permutation Entropy algorithm and to Montserrat-Alvarado González for providing the source code and a detailed description of the SHCC algorithm.

Conflicts of Interest: The authors declare no conflict of interest.

Abbreviations

The following abbreviations are used in this manuscript:

EEG	electroencephalography
BCI	Brain Computer Interfaces
BMI	Brain Machine Interfaces
BNCI	Brain-Neural Computer Interfaces
SNR	Signal to Noise Ratio
CNS	Central Nervous System
AC	Alternating Current
DC	Direct Current
ERP	Event-Related Potential
P300	Positive deflection at 300 ms
ITR	Information Transfer Rate
BTR	Bit Transfer Rate
SIFT	Scale Invariant Feature Transform
SHCC	Slope Horizontal Chain Code
PE	Permutation Entropy
MP	Matching Pursuit
ICU	Intensive Care Unit
EKG	Electrocardiogram
PAA	Period Amplitude Analysis
SVM	Support Vector Machine

References

1. Wolpaw, J.R.; Birbaumer, N.; McFarland, D.J.; Pfurtscheller, G.; Vaughan, T.M. Brain-computer interfaces for communication and control. *Clin. Neurophysiol.* **2002**, *113*, 767–791. [CrossRef]
2. Lutz, W.; Sanderson, W.; Scherbov, S. The coming acceleration of global population ageing. *Nature* **2008**, *451*, 716–719. [CrossRef] [PubMed]
3. Domingo, M.C. An overview of the Internet of Things for people with disabilities. *J. Netw. Comput. Appl.* **2012**, *35*, 584–596. [CrossRef]
4. Guger, C.; Allison, B.Z.; Lebedev, M.A. Introduction. In *Brain Computer Interface Research: A State of the Art Summary 6*; Springer: Cham, Switzerland, 2017; pp. 1–8.
5. Schomer, D.L.; Silva, F.L.D. *Niedermeyer's Electroencephalography: Basic Principles, Clinical Applications, and Related Fields*; Oxford University Press: Oxford, UK, 2017.
6. Puce, A.; Hämäläinen, M.S. A review of issues related to data acquisition and analysis in EEG/MEG studies. *Brain Sci.* **2017**, *7*, 58. [CrossRef] [PubMed]
7. De Vos, M.; Debener, S. Mobile eeg: Towards brain activity monitoring during natural action and cognition. *Int. J. Psychophysiol.* **2014**, *91*, 1–2. [CrossRef] [PubMed]
8. Hartman, A.L. *Atlas of EEG Patterns*; Lippincott Williams & Wilkins: Philadelphia, PA, USA, 2005.
9. Yuste, R.; Goering, S.; Agüeray Arcas, B.; Bi, G.; Carmena, J.M.; Carter, A.; Fins, J.J.; Friesen, P.; Gallant, J.; Huggins, J.E.; et al. Four ethical priorities for neurotechnologies and AI. *Nature* **2017**, *551*, 159–163. [CrossRef] [PubMed]
10. Tzimourta, K.D.; Tsoulos, I.; Bilero, T.; Tzallas, A.T.; Tsipouras, M.G.; Giannakeas, N. Direct Assessment of Alcohol Consumption in Mental State Using Brain Computer Interfaces and Grammatical Evolution. *Inventions* **2018**, *3*, 51. [CrossRef]
11. Kevric, J.; Subasi, A. The Impact of Mspca Signal De-Noising In Real-Time Wireless Brain Computer Interface System. *Southeast Eur. J. Soft Comput.* **2016**, *4*, 1–12. [CrossRef]
12. Cruz, A.; Pires, G.; Nunes, U.J. Double ErrP Detection for Automatic Error Correction in an ERP-Based BCI Speller. *IEEE Trans. Neural Syst. Rehabil. Eng.* **2018**, *26*, 26–36. [CrossRef] [PubMed]
13. Jansen, B.H. Quantitative analysis of electroencephalograms: Is there chaos in the future? *Int. J. Bio-Med. Comput.* **1991**, *27*, 95–123. [CrossRef]

14. Thakor, N.V.; Tong, S. Advances in Quantitative Electroencephalogram Analysis Methods. *Annu. Rev. Biomed. Eng.* **2004**, *6*, 453–495. [CrossRef] [PubMed]

15. Jackson, A.F.; Bolger, D.J. The neurophysiological bases of EEG and EEG measurement: A review for the rest of us. *Psychophysiology* **2014**, *51*, 1061–1071. [CrossRef] [PubMed]

16. Haberman, M.A.; Spinelli, E.M. A multichannel EEG acquisition scheme based on single ended amplifiers and digital DRL. *IEEE Trans. Biomed. Circuits Syst.* **2012**, *6*, 614–618. [CrossRef] [PubMed]

17. Weeda, W.D.; Grasman, R.P.P.P.; Waldorp, L.J.; van de Laar, M.C.; van der Molen, M.W.; Huizenga, H.M. A fast and reliable method for simultaneous waveform, amplitude and latency estimation of single-trial EEG/MEG data. *PLoS ONE* **2012**, *7*, e38292. [CrossRef] [PubMed]

18. Farzan, F.; Atluri, S.; Frehlich, M.; Dhami, P.; Kleffner, K.; Price, R.; Lam, R.W.; Frey, B.N.; Milev, R.; Ravindran, A.; et al. Standardization of electroencephalography for multi-site, multi-platform and multi-investigator studies: Insights from the canadian biomarker integration network in depression. *Sci. Rep.* **2017**, *7*, 7473. [CrossRef] [PubMed]

19. Cole, S.R.; Voytek, B. Brain Oscillations and the Importance of Waveform Shape. *Trends Cogn. Sci.* **2017**, *21*, 137–149. [CrossRef] [PubMed]

20. Buzsáki, G.; Anastassiou, C.A.; Koch, C. The origin of extracellular fields and currents-EEG, ECoG, LFP and spikes. *Nat. Rev. Neurosci.* **2012**, *13*, 407–420. [CrossRef] [PubMed]

21. Giagante, B.; Oddo, S.; Silva, W.; Consalvo, D.; Centurion, E.; D'Alessio, L.; Solis, P.; Salgado, P.; Seoane, E.; Saidon, P.; et al. Clinical-electroencephalogram patterns at seizure onset in patients with hippocampal sclerosis. *Clin. Neurophysiol.* **2003**, *114*, 2286–2293. [CrossRef]

22. Rodenbeck, A.; Binder, R.; Geisler, P.; Danker-Hopfe, H.; Lund, R.; Raschke, F.; Weeß, H.G.; Schulz, H. A review of sleep EEG patterns. Part I: A compilation of amended rules for their visual recognition according to Rechtschaffen and Kales. *Somnologie* **2006**, *10*, 159–175. [CrossRef]

23. Boostani, R.; Karimzadeh, F.; Nami, M. A comparative review on sleep stage classification methods in patients and healthy individuals. *Comput. Methods Programs Biomed.* **2017**, *140*, 77–91. [CrossRef] [PubMed]

24. Dimitriadis, S.I.; Salis, C.; Linden, D. A novel, fast and efficient single-sensor automatic sleep-stage classification based on complementary cross-frequency coupling estimates. *Clin. Neurophysiol.* **2018**, *129*, 815–828. [CrossRef] [PubMed]

25. Uchida, S.; Feinberg, I.; March, J.D.; Atsumi, Y.; Maloney, T. A comparison of period amplitude analysis and FFT power spectral analysis of all-night human sleep EEG. *Physiol. Behav.* **1999**, *67*, 121–131. [CrossRef]

26. Britton, J.W.; Frey, L.C.; Hopp, J.L.; Korb, P.; Koubeissi, M.Z.; Lievens, W.E.; Pestana-Knight, E.M.; St. Louis, E.K. *Electroencephalography (EEG): An Introductory Text and Atlas of Normal and Abnormal Findings in Adults, Children, and Infants*; American Epilepsy Society: Chicago, IL, USA, 2016.

27. Luck, S.J. *An Introduction to the Event-Related Potential Technique*; USA; MIT Press: Cambridge, MA, USA, 2005; Volume 78, p. 388.

28. Tatum, W.; Husain, A.; Benbadis, S.; Kaplan, P. *Handbook of EEG Interpretation*; Demos Medical Publishing: New York, NY, USA, 2008.

29. Cacioppo, J.; Tassinary, L.G.; Berntson, G.G. *The Handbook of Psychophysiology*; Cambridge University Press: Cambridge, UK, 2007.

30. Kappenman, E.S.; Luck, S.J. *The Oxford Handbook of Event-Related Potential Components*; Oxford University Press: Oxford, UK, 2012.

31. Ouyang, G.; Hildebrandt, A.; Sommer, W.; Zhou, C. Exploiting the intra-subject latency variability from single-trial event-related potentials in the P3 time range: A review and comparative evaluation of methods. *Neurosci. Biobehav. Rev.* **2017**, *75*, 1–21. [CrossRef] [PubMed]

32. Sanei, S.; Chambers, J. *EEG Signal Processing*; Wiley: Chichester, UK, 2007.

33. Wulsin, D.F.; Gupta, J.R.; Mani, R.; Blanco, J.A.; Litt, B. Modeling electroencephalography waveforms with semi-supervised deep belief nets: Fast classification and anomaly measurement. *J. Neural Eng.* **2011**, *8*, 036015, [CrossRef] [PubMed]

34. Hirsch, L.J.; Richard, B.P. *Atlas of EEG in Critical Care*; Wiley-Blackwell: Hoboken, NJ, USA, 2010; p. 348.

35. Subha, D.P.; Joseph, P.K.; Acharya U, R.; Lim, C.M. EEG signal analysis: A survey. *J. Med. Syst.* **2010**, *34*, 195–212. [CrossRef] [PubMed]

36. Mak, J.N.; McFarland, D.J.; Vaughan, T.M.; McCane, L.M.; Tsui, P.Z.; Zeitlin, D.J.; Sellers, E.W.; Wolpaw, J.R. EEG correlates of P300-based brain-computer interface (BCI) performance in people with amyotrophic lateral sclerosis. *J. Neural Eng.* **2012**, *9*, 026014. [CrossRef] [PubMed]

37. Müller-Putz, G.R.; Riedl, R.; Wriessnegger, S.C. Electroencephalography (EEG) as a research tool in the information systems discipline: Foundations, measurement, and applications. *Commun. Assoc. Inf. Syst.* **2015**, *37*, 911–948. [CrossRef]

38. Shah, N.A.; Wusthoff, C.J. How to use: Amplitude-integrated EEG (aEEG). *Arch. Dis. Child. Educ. Pract. Ed.* **2015**, *100*, 75–81. [CrossRef] [PubMed]

39. Moher, D.; Liberati, A.; Tetzlaff, J.; Altman, D.G.; Altman, D.; Antes, G.; Atkins, D.; Barbour, V.; Barrowman, N.; Berlin, J.A.; et al. Preferred reporting items for systematic reviews and meta-analyses: The PRISMA statement. *PLoS Med.* **2009**, *6*, e1000097. [CrossRef] [PubMed]

40. Allen, R.L.; Mills, D. *Signal Analysis: Time, Frequency, Scale, and Structure*; John Wiley & Sons: Hoboken, NJ, USA, 2004.

41. Wolpaw, J.; Wolpaw, E.W. *Brain-Computer Interfaces: Principles and Practice*; Oxford University Press: Oxford, UK, 2012; p. 400.

42. Thakor, N. *Quantitative EEG Analysis Methods and Clinical Applications*; Artech House Series Engineering in Medicine and Biology: Norwood, MA, USA, 2009; p. 440.

43. Jaśkowski, P.; Verleger, R. An evaluation of methods for single-trial estimation of P3 latency. *Psychophysiology* **2000**, *37*, 153–162. [CrossRef] [PubMed]

44. Zhang, D.; Luo, Y. The P1 latency of single-trial ERPs estimated by two peak-picking strategies. In Proceedings of the 2011 4th International Conference on Biomedical Engineering and Informatics (BMEI), Shanghai, China, 15–17 October 2011; Volume 2, pp. 882–886.

45. Alvarado-González, M.; Garduño, E.; Bribiesca, E.; Yáñez-Suárez, O.; Medina-Bañuelos, V. P300 Detection Based on EEG Shape Features. *Comput. Math. Methods Med.* **2016**, *2016*, 1–14. [CrossRef] [PubMed]

46. Farwell, L.A.; Donchin, E. Talking off the top of your head: Toward a mental prosthesis utilizing event-related brain potentials. *Electroencephalogr. Clin. Neurophysiol.* **1988**, *70*, 510–523. [CrossRef]

47. Klein, F.F. A waveform analyzer applied to the human EEG. *IEEE Trans. Biomed. Eng.* **1976**, *23*, 246–252. [CrossRef] [PubMed]

48. Yamaguchi, T.; Fujio, M.; Inoue, K.; Pfurtscheller, G. Design Method of Morphological Structural Function for Pattern Recognition of EEG Signals During Motor Imagery and Cognition. In Proceedings of the 2009 Fourth International Conference on Innovative Computing, Information and Control (ICICIC), Kaohsiung, Taiwan, 7–9 December 2009; pp. 1558–1561.

49. Uchida, S.; Matsuura, M.; Ogata, S.; Yamamoto, T.; Aikawa, N. Computerization of Fujimori's method of waveform recognition a review and methodological considerations for its application to all-night sleep EEG. *J. Neurosci. Methods* **1996**, *64*, 1–12. [CrossRef]

50. Vincent, E.; Deville, Y. *Handbook of Blind Source Separation—Independent Component Analysis and Applications*; Elsevier: Amsterdam, The Netherlands, 2010; p. 831.

51. Mallat, S.G.; Zhang, Z. Matching Pursuits With Time-Frequency Dictionaries. *IEEE Trans. Signal Process.* **1993**, *41*, 3397–3415. [CrossRef]

52. KS, S.C.; Mishra, A.; Shirhatti, V.; Ray, S. Comparison of Matching Pursuit Algorithm with Other Signal Processing Techniques for Computation of the Time-Frequency Power Spectrum of Brain Signals. *J. Neurosci.* **2016**, *36*, 3399–3408.

53. Cohen, M.X. *Analyzing Neural Time Series Data: Theory and Practice*; MIT Press: Cambridge MA, USA, 2014.

54. Vařeka, L. Matching pursuit for p300-based brain-computer interfaces. In Proceedings of the 2012 35th International Conference on Telecommunications and Signal Processing, Prague, Czech Republic, 3–4 July 2012; pp. 513–516.

55. Bandt, C.; Pompe, B. Permutation Entropy: A Natural Complexity Measure for Time Series. *Phys. Rev. Lett.* **2002**, *88*, 174102. [CrossRef] [PubMed]

56. Nicolaou, N.; Georgiou, J. Permutation entropy: A new feature for brain-computer interfaces. In Proceedings of the 2010 IEEE Biomedical Circuits and Systems Conference, BioCAS 2010, Paphos, Cyprus, 3–5 November 2010; pp. 49–52.

57. Keller, K.; Mangold, T.; Stolz, I.; Werner, J. Permutation Entropy: New Ideas and Challenges. *Entropy* **2017**, *19*, 134. [CrossRef]

58. Unakafova, V.; Keller, K. Efficiently Measuring Complexity on the Basis of Real-World Data. *Entropy* **2013**, *15*, 4392–4415. [CrossRef]

59. Berger, S.; Schneider, G.; Kochs, E.F.; Jordan, D. Permutation entropy: Too complex a measure for EEG time series? *Entropy* **2017**, *19*, 692. [CrossRef]

60. Lowe, G. SIFT—The Scale Invariant Feature Transform. *Int. J. Comput. Vis.* **2004**, *2*, 91–110. [CrossRef]

61. Edelman, S.; Intrator, N.; Poggio, T. Complex cells and object recognition. 1997, unpublished manuscript.

62. Ramele, R.; Villar, A.J.; Santos, J.M. BCI classification based on signal plots and SIFT descriptors. In Proceedings of the 4th International Winter Conference on Brain-Computer Interface (BCI), Yongpyong, South, 22–24 February 2016; pp. 1–4.

63. Zhang, R.; Xu, P.; Guo, L.; Zhang, Y.; Li, P.; Yao, D. Z-Score Linear Discriminant Analysis for EEG Based Brain-Computer Interfaces. *PLoS ONE* **2013**, *8*, e74433. [CrossRef] [PubMed]

64. Vedaldi, A.; Fulkerson, B. VLFeat—An open and portable library of computer vision algorithms. *Design* **2010**, *3*, 1–4.

65. Rakotomamonjy, A.; Guigue, V. BCI Competition III: Dataset II—Ensemble of SVMs for BCI P300 Speller. *IEEE Trans. Biomed. Eng.* **2008**, *55*, 1147–1154. [CrossRef] [PubMed]

66. Chavarriaga, R.; Fried-Oken, M.; Kleih, S.; Lotte, F.; Scherer, R. Heading for new shores! Overcoming pitfalls in BCI design. *Brain-Comput. Interfaces* **2017**, *4*, 60–73. [CrossRef] [PubMed]

67. Quiroga, R.Q.; Garcia, H. Single-trial event-related potentials with wavelet denoising. *Clin. Neurophysiol.* **2003**, *114*, 376–390. [CrossRef]

68. Renard, Y.; Lotte, F.; Gibert, G.; Congedo, M.; Maby, E.; Delannoy, V.; Bertrand, O.; Lécuyer, A. OpenViBE: An Open-Source Software Platform to Design, Test, and Use Brain Computer Interfaces in Real and Virtual Environments. *Presence Teleoper. Virtual Environ.* **2010**, *19*, 35–53. [CrossRef]

69. Riccio, A.; Simione, L.; Schettini, F.; Pizzimenti, A.; Inghilleri, M.; Belardinelli, M.O.; Mattia, D.; Cincotti, F. Attention and P300-based BCI performance in people with amyotrophic lateral sclerosis. *Front. Hum. Neurosci.* **2013**, *7*, 732. [CrossRef] [PubMed]

70. Brunner, C.; Blankertz, B.; Cincotti, F.; Kübler, A.; Mattia, D.; Miralles, F.; Nijholt, A.; Otal, B. BNCI Horizon 2020—Towards a Roadmap for Brain/Neural Computer Interaction. *Lect. Notes Comput. Sci.* **2014**, *8513*, 475–486.

71. Rao, R.P.N. *Brain-Computer Interfacing: An Introduction*; Cambridge University Press: New York, NY, USA, 2013; p. 319.

72. Clerc, M.; Bougrain, L.; Lotte, F. *Brain-Computer Interfaces, Technology and Applications 2 (Cognitive Science)*; ISTE Ltd. and Wiley: London, UK, 2016; p. 324.

73. Gramfort, A.; Luessi, M.; Larson, E.; Engemann, D.A.; Strohmeier, D.; Brodbeck, C.; Goj, R.; Jas, M.; Brooks, T.; Parkkonen, L.; et al. MEG and EEG data analysis with MNE-Python. *Front. Neurosci.* **2013**, *7*, 267. [CrossRef] [PubMed]

74. Da Pelo, P.; De Tommaso, M.; Monaco, A.; Stramaglia, S.; Bellotti, R.; Tangaro, S. Trial latencies estimation of event-related potentials in EEG by means of genetic algorithms. *J. Neural Eng.* **2018**, *15*, 026016. [CrossRef] [PubMed]

75. Harris, F.J. On the Use of Windows for Harmonic Analysis with the Discrete Fourier Transform. *Proc. IEEE* **1978**, *66*, 51–83. [CrossRef]

76. Blankertz, B. Documentation Second Wadsworth BCI Dataset (P300 Evoked Potentials) Data Acquired Using BCI2000 P300 Speller Paradigm. BCI Classification Contest November. Available online: http://www.bbci.de/competition/ii/albany_desc/albany_desc_ii.pdf (accessed on 14 November 2018).

77. Krusienski, D.J.; Sellers, E.W.; Cabestaing, F.; Bayoudh, S.; McFarland, D.J.; Vaughan, T.M.; Wolpaw, J.R. A comparison of classification techniques for the P300 Speller. *J. Neural Eng.* **2006**, *3*, 299–305. [CrossRef] [PubMed]

78. Kaper, M.; Meinicke, P.; Grossekathoefer, U.; Lingner, T.; Ritter, H. BCI competition 2003—Data set IIb: Support vector machines for the P300 speller paradigm. *IEEE Trans. Biomed. Eng.* **2004**, *51*, 1073–1076. [CrossRef] [PubMed]

79. Boiman, O.; Shechtman, E.; Irani, M. In defense of nearest-neighbor based image classification. In Proceedings of the 26th IEEE Conference on Computer Vision and Pattern Recognition (CVPR), Anchorage, AK, USA, 23–28 June 2008.

Brain Sci. **2018**, *8*, 199

80. Käthner, I.; Halder, S.; Hintermüller, C.; Espinosa, A.; Guger, C.; Miralles, F.; Vargiu, E.; Dauwalder, S.; Rafael-Palou, X.; Solà, M.; et al. A multifunctional brain-computer interface intended for home use: An evaluation with healthy participants and potential end users with dry and gel-based electrodes. *Front. Neurosci.* **2017**, *11*, 286. [CrossRef] [PubMed]

81. Neuper, C.; Müller, G.; Kübler, A.; Birbaumer, N.; Pfurtscheller, G. Clinical application of an EEG-based brain–computer interface: A case study in a patient with severe motor impairment. *Clin. Neurophysiol.* **2003**, *114*, 399–409. [CrossRef]

82. Kundu, S.; Ari, S. P300 Detection with Brain–Computer Interface Application Using PCA and Ensemble of Weighted SVMs. *IETE J. Res.* **2018**, *64*, 406–414. [CrossRef]

83. Gribonval, R.; Rauhut, H.; Schnass, K.; Vandergheynst, P. Atoms of all channels, unite! Average case analysis of multi-channel sparse recovery using greedy algorithms. *J. Fourier Anal. Appl.* **2008**, *14*, 655–687. [CrossRef]

84. Bragg, M.J. The Challenge of Crafting Intelligible Intelligence. In Proceedings of the ACM Symposium on User Interface Software and Technology UIST 2018, Berlin, Germany, 14–17 October 2018.

85. Skoog, D.A.; West, D.M.; Holler, F.J.; Crouch, S.R. *Analytical Chemistry: An Introduction*; Saunders College Publishing: Philadelphia, PA, USA, 2000.

86. Owens, T.J.; Zandt, G.; Taylor, S.R. Seismic evidence for an ancient rift beneath the Cumberland Plateau, Tennessee: A detailed analysis of broadband teleseismic P waveforms. *J. Geophys. Res. Solid Earth* **1984**, *89*, 7783–7795. [CrossRef]

87. Stockman, G.; Kanal, L.; Kyle, M. Structural pattern recognition of carotid pulse waves using a general waveform parsing system. *Commun. ACM* **1976**, *19*, 688–695. [CrossRef]

brain
sciences

MDPI

Article

A Motivational Model of BCI-Controlled Heuristic Search

Marc Cavazza

Department of Computing and Information Systems, University of Greenwich, London SE10 9LS, UK;
m.cavazza@gre.ac.uk; Tel.: +44-20-8331-8512; Fax: +44-20-8331-8665

Received: 11 June 2018; Accepted: 17 August 2018; Published: 31 August 2018

Abstract: Several researchers have proposed a new application for human augmentation, which is to provide human supervision to autonomous artificial intelligence (AI) systems. In this paper, we introduce a framework to implement this proposal, which consists of using Brain–Computer Interfaces (BCI) to influence AI computation via some of their core algorithmic components, such as heuristic search. Our framework is based on a joint analysis of philosophical proposals characterising the behaviour of autonomous AI systems and recent research in cognitive neuroscience that support the design of appropriate BCI. Our framework is defined as a motivational approach, which, on the AI side, influences the shape of the solution produced by heuristic search using a BCI motivational signal reflecting the user's disposition towards the anticipated result. The actual mapping is based on a measure of prefrontal asymmetry, which is translated into a non-admissible variant of the heuristic function. Finally, we discuss results from a proof-of-concept experiment using functional near-infrared spectroscopy (fNIRS) to capture prefrontal asymmetry and control the progression of AI computation of traditional heuristic search problems.

Keywords: augmented cognition; brain–computer interfaces; superintelligence; heuristic search

---

1. Introduction and Rationale

Human augmentation aims at extending human cognitive abilities, often in a situated, task-specific fashion. Previous research has demonstrated through various implemented prototypes and experiments the feasibility of extending human perceptive abilities or information processing and decision-making abilities [1,2]. In the latter case, Artificial Intelligence (AI) techniques are poised to play a significant role in providing the task-specific information processing power supporting the augmentation aspects. A constant feature, and a defining aspect of human augmentation, is that locus of control remains strictly with the human, and the human task dynamics is left largely unchanged. The information processing ability provided by the augmenting system is inserted into the natural human activity, in a user-centred way, largely like augmented reality systems enhance world perception through advanced imaging abilities. One such example is cortically coupled perception [3], in which user active analysis of satellite images is augmented by the EEG-based detection of perceptive signals: in this experimental system, the human analyst approach to image exploration is essentially unchanged.

Although human augmentation systems have been developed prior to the popularisation of Brain–Computer Interfaces (BCI), these have taken a more prominent role in recent years, as they offer a seamless mechanism to capture elements of human cognitive processes in a way that enables the synchronisation of computations [1,2]. With the rise of autonomous intelligent systems, a new application of human augmentation has been suggested in order to keep humans in control of autonomous AI systems whose performance could potentially exceed even that of human experts.

After years of inflated expectations about AI, recent progress, primarily in machine learning, has led to much-advertised successes [4,5] and renewed confidence in AI advances. Paradoxically, this situation has also fuelled the preoccupation that AI progress will eventually constitute a threat to the well-being of humans. Researchers across a variety of disciplines have taken the stage to forewarn of the potential adverse consequences of unregulated AI progress, amongst which the automation of white-collar jobs [6], the development of AI-endowed autonomous warfare [7], or even the rise of superintelligent AI entities [8,9]. Whether or not this superintelligence threat will materialise, the shorter-term availability of advanced AI systems able to outperform human experts at an increasing number of professional tasks is sufficient to justify research into hybrid cognitive systems.

The imbalance between humans and AI systems stems largely from the inability of humans to engage with, even less control, the automatic reasoning mechanisms underpinning AI systems. This stems largely from the scale and pace of data processing, which is not compatible with the timeline of human decision making. It can also be noted that this lack of surveyability is not strictly attributable to a representational issue (e.g., sub-symbolic versus symbolic), as complex search systems, including statistical ones such as Monte Carlo tree search [5], remain based on the discrete step granularity of search.

There is thus a case for additional research exploring a synergy between humans and AI systems, which should aim at endowing humans with high-level control abilities sufficient to steer the flow of AI computation, irrespective of its low-level details, while preserving an understanding of the computation goal. Several authors have specifically suggested human augmentation as a potential solution to the threat posed by superintelligence, augmentation being often achieved through BCI implementations. Although most of these proposals remain largely underspecified, and some are not always consistent with the state-of-the-art of BCI systems, it is worth briefly reviewing the commonalities between them. Bostrom dedicates a section of his book [9] (p. 169) to the potential of BCI for controlling superintelligence: however, his analysis is moderately optimistic, largely because he equates BCI with its invasive implementations (depth electrodes or electrocorticography (ECoG)), and raises legitimate concerns about acceptance, maturity of the technology, side effects, and user safety. Kennedy [10] suggests BCI-based augmentation primarily as an alternative pathway to autonomous superintelligence rather than as a control mechanism, and rightly identifies BCI signal/information bandwidth as a major challenge. Skulimowski reviews several candidate scenarios for superintelligence [11], one of which involves human control through BCI connection. Finally, Barrett and Baum, in their review of pathways to (artificial) superintelligence [12], discuss several risk reduction interventions, one of which includes human augmentation through BCI.

Despite being initially identified as human augmentation, it would actually imply a paradigm shift, because the main information processing cycle would be driven by the autonomous AI rather than by the human, as is customary in traditional cognitive augmentation systems; here, the human user would be included in a supervisory capacity. To be successfully implemented, this framework should not require a transformation of the AI technology to support user intervention (e.g., mixed initiative), as it might compromise efficiency and the very advantage of autonomy. The challenge we are addressing here consists precisely in providing minimally invasive supervision by the human user. To summarise previous literature, the rationale for providing supervisory control can be described from two complementary perspectives: (i) controlling the nature of the solution during its calculation (in terms of optimality, solution parameters, or other application-related criteria), and (ii) ensuring compliance with ethical standards.

In this paper, we introduce a candidate framework aimed at controlling the behaviour of autonomous AI systems using a BCI. This unique combination of BCI and AI is meant to integrate BCI input directly at the level of AI algorithmic computation so as to influence inner mechanisms in a principled manner, being compatible with typical BCI information bandwidth and without imposing additional restrictions on the nature of AI computation.

Although the current approach shares important aspects with augmented cognition, it differs fundamentally by the fact that the main computation is actually determined by autonomous AI mechanisms with the user supervising the computation rather than actually driving the task, as in cortically coupled vision or enhanced information retrieval [2]. The combination of the user and the AI system forms a hybrid cognitive system in which some high-level executive control would be retained by the human, while autonomous AI would form the main cognitive process. To account for the complex spectrum of human–system integration, previous literature has used terminology such as symbiotic systems [13] or human–computer confluence [14], and we should refer to our approach as BCI-controlled heuristic search, categorising the type of system we are aiming for as a hybrid cognitive system.

One of the main objectives is to achieve consistency between user intentions and the principles that can affect the progression of AI computation: to that effect, we will review several principles that reconcile active BCI, user cognitive mechanisms, and AI computation dynamics. In the next sections, after introducing the issues emerging from autonomous AI systems and reviewing relevant BCI augmentation systems, we discuss basic AI mechanisms (i.e., heuristic search) that can serve as a target for BCI influence. We then explore cognitive processes that could be harnessed to provide control over AI computation. We will emphasise cognitive mechanisms around motivation, which range from reward expectation to risk propensity, trying to relate them to compatible concepts that characterise the progress of AI computations in terms of result anticipation. Even before being fully fledged from a theoretical perspective, this framework has been the object of early proof-of-concept testing through a fully implemented prototype, whose results are briefly analysed as additional input into the proposed framework. Finally, we take a system design perspective to review the conditions for a successful implementation of the framework, as well as possible implementation variants.

2. A Motivational Model of AI Control

Theoretical research on superintelligence has suggested various approaches and mechanisms to ensure it will stay under human control. In the first instance, we will consider that, from a technical perspective, the human augmentation mechanisms proposed for superintelligent systems should not fundamentally differ from those to be associated with shorter-term autonomous AI systems endowed with advanced planning, decision making, or information analysis abilities. Bostrom has advocated one specific control mechanism, which he characterises as motivation selection methods [9] (p. 169), or methods that would shape the nature of the solution produced by the AI system. While his original discussion is influenced by a rather anthropomorphic view of the AI's goals and intentions, this philosophy can be extended to more technical visions of AI systems to describe the type of solution produced, whether this type is defined in terms of goal properties or solution properties (when the shape of the solution, seen as a sequence of actions towards the goal, constitutes a desirable property of the output). For instance, instead of indirect normativity [9] (p. 169) influencing the set of values used by the AI in the pursuit of a solution, the nature of a solution could be shaped by the user according to shared concepts characterising the nature of the solution. Candidate concepts would include reward anticipation, risk taking, and solution optimality: we shall develop in the forthcoming sections how these concepts can be related to cognitive motivational dimensions and how they can be made accessible to BCI input. In the next section, we will first lay out some AI basic mechanisms that rest at the heart of many AI systems and can constitute a target for the user-based influence of AI computation.

In the above model, motivation has been defined primarily in relation to goal setting and goal pursuit. Recent research in cognitive neuroscience [15] uses a compatible definition of motivation that can be made interoperable with AI technology concepts. In addition, it identifies the involvement of specific brain regions in a way that supports the design of appropriate BCI. From a cognitive perspective as well, motivation is conceived of as being goal directed [16,17]. The relationship to the goal has been further refined into planning and implementing stages [18], also suggesting that goal setting is primarily motivational, while goal striving is best characterised in terms of volitional factors [18]. According to [16], the neural systems implicated in the internal representation of cognitive goals overlap significantly with those dealing with the generation of motivated behaviours. In particular, the lateral prefrontal cortex (PFC) might serve as a convergence zone in which motivational and cognitive variables are integrated [16].

The identification of specific brain regions whose activation may reflect motivational dimensions is an essential step in designing appropriate BCI. In terms of activity measurement, there is a substantial body of work associating PFC asymmetry with motivational direction [19], which originates with the study of approach/withdrawal as a motivational dimension [20]. This research has pioneered the measurement of prefrontal asymmetry using EEG signals [20], left asymmetry being associated with the expression of approach.

The relationship of frontal EEG asymmetry with motivational variables has been recently reviewed by Smith et al. [21] and Harmon-Jones and Gable [17], who have related resting left prefrontal asymmetry to individual differences in self-reported trait approach motivation. In addition, they have found this relationship to be stronger in the context of incentive anticipation. Moreover, there are strong relations between motivation and reward anticipation: for instance, lateral PFC activation is modulated by the level of reward offered [22,23]. Amodio et al. [24] have analysed the correlates of PFC asymmetry from a regulatory perspective. More specifically, they found approach regulation to be most relevant to "pre-goal states", during which efforts are mobilised towards the goal. This needs to be reanalysed from the prism of a hybrid cognitive system, which could involve a mix of goal setting and goal pursuit depending on the information visible to the user from the AI computation but, in any case, is compatible with a mediation from prefrontal asymmetry.

Prefrontal asymmetry as a marker of approach [17] has been extensively studied by EEG under three different conditions: (i) at rest, (ii) as a dynamic response to a cognitive situation or an affective stimulus, and (iii) under volitional control through neurofeedback (NF). To understand the dynamics of prefrontal asymmetry, it is worth noting that its value is determined approximately for half by its resting value (trait) and for another half by its dynamic value (state): this is in particular what makes it amenable to volitional control through NF, although the trait component may introduce ceiling effects rendering some subjects more prone to dynamic changes than others.

Functional Magnetic Resonance Imaging (fMRI) studies have been carried out to uncover the anatomical basis of prefrontal asymmetry in the context of motivational phenomena [25,26]. In addition, real-time-fMRI (rt-fMRI) experiments have demonstrated the controllability of prefrontal asymmetry [27] including comparisons to EEG-based NF. Functional Near Infrared Spectroscopy (fNIRS) studies of the PFC have been dedicated to affective interaction [28] as well as executive control. fNIRS is also amenable to NF implementation that supports BCI, and we have successfully used it for BCI-based prefrontal symmetry in a context of distinguishing approach from valence [29]. Harmon-Jones et al. [19] have questioned the exclusive role of the dorsolateral prefrontal cortex (DLPFC) in accounting for BCI signals for approach, on the basis that EEG and metabolic methods, such as fMRI, measure different activities for different cellular populations [30], while noting that EEG findings were still corroborated by experiments with transcranial stimulation (see for instance, [31]). In their most recent review, Harmon-Jones and Gable [17] have considered that fMRI may actually show more complex patterns of activation without this invalidating the central role of DLPFC and the use of EEG to measure prefrontal asymmetry.

When placing the human user in a position of high-level arbitration of autonomous AI computations, it is tempting to resort to a metaphor of executive control, even more so when resorting to neural signals originating in the prefrontal cortex. A hybrid model of executive control could be envisioned, by redefining executive control for a hybrid cognitive system comprised of the human and the autonomous AI, in which human executive control would apply to the deliberative AI part instead of the human part. One specific question arising when considering cognitive control in the context of hybrid cognitive systems is the extent to which prefrontal cognitive control mechanisms that have been described to operate on internal cognitive mechanisms would apply to hybrid control situations where the generation of hypotheses, or anticipation of rewards, is actually not the result of the human cognitive processes but of their appraisal of the AI calculation progress.

Smith et al. [21] have actually related EEG prefrontal asymmetry not just to motivation but also executive functions. Current integrative models of executive function control [32] distinguish between hot (affective) and cold (deliberative) executive control and tend to associate the DLPFC with cold control and the orbitofrontal cortex (OFC) with motivation and reward anticipation. This would be consistent with source-localisation studies, which have suggested that frontal EEG asymmetry at rest is mediated by left DLPFC and OFC activation [33].

Yet, Auperle et al. [22] (following, amongst others, [34,35]) have distinguished specific roles for OFC and DLPFC. They suggest that OFC is involved in determining the value of rewards, while DLPFC incorporates these values when planning for the execution of a decision or response. Compatible findings had been reported by Tanaka et al. [36], with OFC involved in learning from the present state and DLPFC in learning from predictable future states. The original work from Wallis and Miller [37], based on a primate model, established that OFC encoded the reward value alone, while DLPFC encoded both the reward value and the forthcoming response. Li et al. [38] have also suggested that subjects could use the DLPFC to dynamically adjust outcome responses depending on the usefulness of action-outcome information, implying that they could make use of instructed knowledge rather than simply trial and error outcomes. The role of the left PFC has been described from a hierarchical perspective alongside a rostro-caudal hierarchy as introduced by Coutlee and Huettel [39]. In that context, the DLPFC, whose activity has the prominent role in PFC asymmetry, is considered to be involved in "mid-level abstraction control", which would be compatible with the goal-oriented role discussed above.

Despite the overlap between motivational and cognitive factors in the PFC, it is difficult to conclude that hybrid cognitive systems could implement executive control simply by transposing human cognitive mechanisms and dissociating human executive control from other cognitive processes, the latter being substituted with an AI system, without a better understanding of the actual control signals and required information bandwidth. We should then entrust control of the hybrid system to the motivational component, whose signal properties and cognitive activation are better understood, without ruling out that in the context of observing the progression of AI computations, these may still interfere in part with executive functions. While some details of the framework remain to be refined—in particular, the exact balance between goal definition and goal pursuit—the above discussion contains sufficient evidence of the appropriateness of a motivational framework to support the interactive component of a hybrid cognitive system.

2.1. Heurisitc Search in AI Control

Implementing cognitive control over AI computation requires the identification of a target computational element, which is generic enough to support one of more classes of AI systems and would not require altering the nature of AI calculations themselves to deploy explicit control mechanisms. One possible target mechanism would consist of the basic elements of AI computation, such as search. The underlying hypothesis is that altering the basic component of heuristic search offers significant leverage on the behaviour of the entire AI computation that derives from it. Of all the algorithmic components underpinning the implementation of AI systems, heuristic search enjoys a central position and also one that has persisted from the early days of AI problem solving to the most recent successes of AI technology.

Heuristic search is in itself a problem-solving technique supporting direct resolution for puzzles such as the Rubik's cube [40] or spatialised optimisation problems such as the travelling salesman problem or equivalent problems [41]. It has been embedded in a large range of real-world AI applications, from speech recognition to sequence alignment in bioinformatics and many others [42]. However, its real power derives from its incorporation in complex problem-solving techniques supporting more sophisticated knowledge representation, such as search-based planning, which has become the dominant planning technique [43], or question answering systems [44] of the type popularised by IBM's Watson™. Within these systems, modifications of the basic search mechanisms are potentially able to affect the generation of solutions in the application's semantic domain without the requirement for domain control knowledge.

The most generic mechanism to influence heuristic search is to act upon the heuristic itself: in particular, this mechanism can leverage on search progression in order to influence higher-level AI computations, without requiring ad hoc or application knowledge. The formal properties of heuristic functions have been extensively studied, and as a consequence, the effects of some heuristics' modification are well understood, and their mathematical properties established. For instance, the behaviour of the entire search process towards an optimal solution is determined by the admissible nature of the heuristic function [45]. It has subsequently been demonstrated that heuristic functions departing from admissibility could be used to trade solution optimality for computational speed (Figure 1). More importantly, it has been established that this departure from optimality could be limited while still showing beneficial effects on search performance: this is referred to as bounded non-admissibility [45] or sometimes ε-admissibility. The main mechanisms to design non-admissible heuristics include dynamic weighting of the heuristic function and focal search. Dynamic weighting consists of allocating different weights to the cost function (g) and the heuristic estimate function (h), generally increasing the latter's weighting above 0.5. One early implementation of dynamic weighting is Pohl's depth-dependent dynamic weighting [46], whose rationale is to increase the role of the heuristic component as search progresses towards the goal. This was one of the early demonstrations that heuristics could be modified during the search process itself, such dynamic modifications entailing interactive approaches without requiring the transformation of the baseline search algorithm into a real-time version. The generic approach to dynamic weighting is defined without consideration of search depth simply as a weighted formula for the evaluation function [47], where n is the node considered, g the cost function, h the heuristic estimation function, and ω the weighting coefficient:

$$f(n) = (1 - \omega) \times g(n) + \omega \times h(n). \tag{1}$$

It has been established that dynamic weighting results in the heuristic not being admissible, such non-admissibility being, however, bounded as a function of the weighting coefficient [48] (which makes dynamic weighting approaches ε-admissible).

Another major non-admissible search paradigm is known as focal search. It consists of applying a secondary heuristic to refine the selection of the most promising nodes selected by the main heuristic and one of its early descriptions is A*ε [45] (p. 89). The underlying mechanism consists of creating a subset of the most promising nodes under consideration (the OPEN list), this subset being called FOCAL. Instead of selecting the best node from OPEN, the search algorithm will select the best node from FOCAL using the secondary heuristic function to that purpose. The canonical description of FOCAL search has been demonstrated to be ε-admissible, as long as the size of FOCAL is limited, and is actually referred to as A*ε. It should not be confused with methods for combining multiple primary heuristics, which, unlike A*ε, still operate on the original node selection mechanism within the OPEN list. The original description of A*ε made explicit reference to minimising computational costs [45] (p. 89), hence again trading optimality for computational speed, and the secondary heuristic was based on such a computational cost estimate. However, there is no a priori restriction on the nature of the secondary heuristic. In particular, it can be used to incorporate application semantics into the search process, for instance, through the evaluation of specific state configurations. Although these mechanisms have been described as studies of the fundamental properties of heuristic search, they have also generated real-world applications: for instance, a recent implementation of FOCAL search has been used for Unmanned Aerial Vehicles (UAV) coordination through the enhanced conflict-based search mechanism [49].

From a hybrid cognitive system perspective, the mechanisms underpinning non-admissibility can constitute appropriate targets for intervention, provided the behaviour of non-admissible variants can be attributed cognitive significance by the user within the motivational framework outlined above. The first illustration of the latter point would be the concept of speed–accuracy trade-off, which is the cornerstone of ε-admissible search [45], and has also been explicitly identified in the cognitive literature on motivation–cognition interaction [50].

An alternative cognitive interpretation would be to consider risk as a unifying concept. From the AI perspective, departing from admissibility carries the risk of producing a solution whose cost is higher than that of the optimal solution [51]. From a cognitive neuroscience perspective, this would be based on several findings on the correlation of resting PFC asymmetry with sensation seeking and risk acceptance [52] or the effect of transcranial direct current stimulation-induced right PFC suppression on risk acceptance [31].

However, on the AI side, non-admissibility actually corresponds to bounded risks (i.e., somehow acceptable by nature) and, in many cases, deviations from the optimal solution are actually rather minimal. There is no proper quantification of risk in non-admissible heuristic search unlike the case with explicit risk-based approaches to search such as R_δ* [45]. This algorithm bases node expansion on the error probability distribution for the heuristic function, thereby formalising the risk of ignoring a more promising direction in search than the one taken. Even so, insofar as such risk may not be clearly visible to subjects during interaction, it is unlikely to redefine the motivational framework as a risk-based approach because of the lack of conscious exposure to risk taking.

The basic mechanism we have chosen for influencing heuristic search progression through bounded non-admissibility consists of dynamic weighting of the A* heuristic function. There is significant background work on dynamic weighting from Pohl's early work on depth-dependent dynamic weighting [46] to recent anytime variants of A*, in which the same algorithm is run repeatedly starting with the more suboptimal solution (highest weighting) [53].

We use a standard but complete A* implementation [45] (p. 75), which has been modified to incorporate dynamic weighting of its evaluation function, resulting in a weighted A* (WA*) implementation. In this version, dynamic weighting can take place from the onset of the search or be triggered once a certain percentage of the search space has been explored (for pre-set configurations such as the 8-puzzle). This version has supported both preliminary tests, which were dedicated to study search progression so as to determine how best to influence it (Figures 1 and 2), and the actual BCI hybrid search experiments reported below.

Although heuristic search algorithms can be applied to a wide range of problems, the actual time dynamics of the search process differ significantly across problems, initial conditions, and search methods. A given search problem can be characterised by the shape of its search space, and the extent to which progression towards the solution is monotonic or requires extensive backtracking.

For preliminary tests characterising solution progression, we have run various search problems from a database of previously resolved 8-puzzle configurations [54], which identify search problems (8-puzzle initial and goal configuration) in terms of number of solutions or solution length. Figure 1a shows the reduction in search space for various values of the WA* weighting coefficient, while Figure 1b illustrates the influence of intervention timing: the actual variation in search space for a fixed weighting coefficient depends on the stage of search progression at which dynamic weighting is applied.

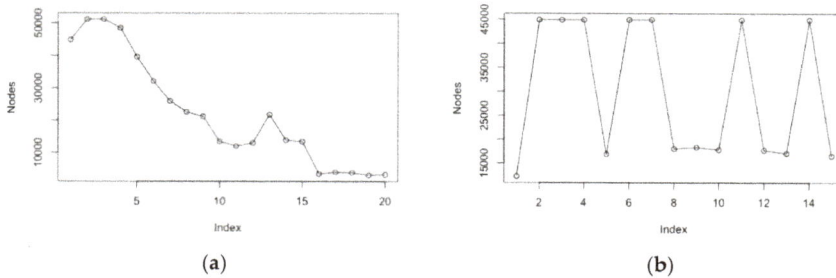

Figure 1. The speed–accuracy trade-off (impact of weighted A* (WA*) on heuristic search performance), illustrated through the reduction in the number of nodes explored to reach a solution. (**a**) Reduction in search space for the 8-puzzle depending on the variation of the weighting coefficient (*x* axis, arbitrary units) (**b**) Restriction of impact depending on the stage of intervention for the 8-puzzle, for a weighting coefficient (0.57) known to reduce the search space.

In addition, some search problems place greater emphasis on the shape of the solution than on simply reaching the goal state. In this context, the impact of shifting to a non-admissible heuristic search would depend on the nature of the search progression itself (i.e., reducing the amount of backtracking or accelerating the monotonic progression towards the solution). It is thus necessary to confirm the ability of the bounded non-admissible search to improve search progression for different problem configurations.

Figure 2a shows the variation of the heuristic value during an 8-puzzle solution for a configuration known to have a 30+ move solution [54], with significant oscillations of the heuristic value indicative of extensive backtracking. Figure 2b shows the variation of the heuristic value for the same problem when shifting to a non-admissible search with a weighting value of 0.55 for the heuristic function.

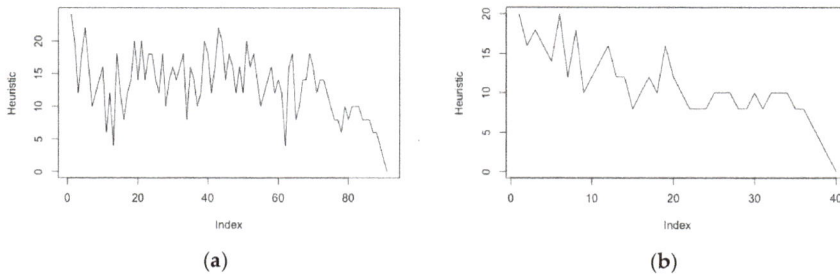

Figure 2. The impact of dynamic weighting on solution progression and backtracking. For the same configuration of the 8-puzzle (**a**) shows significant backtracking with a default heuristic function (A*), while in (**b**) WA* with fixed 0.575 weighting shows a more monotonic progression towards the goal state (as well as a faster computation).

2.2. The Integration Challenge

Integrating control of AI computations assumes a number of conditions for the implementation of the hybrid cognitive system framework. Firstly, the neural signal should be quantified, and its variation range should support a mapping onto defined parameters of the AI calculation. Such grounding can be found in the statistical correlations encountered in previous works, as well as the known magnitudes of signal variations above a baseline. For instance, when considering prefrontal asymmetry in a motivational framework, the situation should be differentiated between EEG and metabolic signals (fMRI, fNIRS), because the latter do not benefit from a fixed prefrontal asymmetry baseline, unlike the trait property of EEG prefrontal asymmetry [17].

Secondly, the signal should be controllable by subjects, implicitly or through an explicit cognitive strategy. The challenge here is to train subjects in developing cognitive strategies that are as specific to the task considered as possible and do not use confounding signals. For instance, prefrontal asymmetry is often influenced by valence in addition to approach, which explains that positively valenced cognitive strategies, such as personal autobiographic memories, can be successful in sustaining the BCI signal [27,55]. However, such cognitive strategies risk being distractive and are decorrelated from the observation of AI computation progress: this could constitute a case for NF training, which is generally reserved for clinical rather than user interface applications. The increase in NF performance, which is generally observed after a few training sessions, could support implicit, non-distracting cognitive strategies. Finally, the users should be responding to a real-time presentation of the progression of AI calculation so that their intervention is relevant in terms of influencing it. Several visualisation strategies will be introduced in the next sections.

Volitional control should be implemented through BCI input supported by specific user training and cognitive strategies. Most literature using prefrontal asymmetry as an active BCI signal has implemented a NF paradigm, most probably because it sought inspiration from the significant literature on PFC asymmetry NF for clinical applications [27,56,57]. In this context, the user intervention can be best described as a motivational response targeting the current evolution of the AI computation.

The integration process at the heart of BCI-controlled search relies on two main dimensions. The first one is the nature of the feedback signal used to convey a sense of search progression and direction: by giving the users a sense of how the search is progressing, it enables them to react accordingly on either time progression or, when available, the nature of the solution most likely to emerge. The second one is the temporal aspects of user intervention, which can be subdivided into timing and frequency of intervention. Timing refers to the time relation between user intervention and the overall duration of the AI computation: it is generally made possible by the extended nature of AI computations. When the progression of the solution can be conveyed meaningfully to the user, this may create the opportunity to guide the search process at various stages assuming again that the duration of the computation is significantly longer than the BCI epochs required for input. The repetition of interventions would then define a frequency of user interventions.

In the next section, we review several options for implementing the above dimensions and how they can be combined to implement various BCI search paradigms.

2.3. Intervention and Search Dynamics

It may seem a paradox to suggest a mechanism for interacting with an offline heuristic search algorithm, considering that there is no shortage of real-time variants of A*. However, there is a long history of repeatedly running heuristic search algorithms with modified heuristics to speed up the remainder of the computation, which was at the heart of various "anytime" variants of A* [48]. Making the search process responsive at specific progression intervals differs from the real-time heuristic search philosophy (e.g., RTA* [58]) in terms of heuristic value calculation (depth-bound lookahead versus goal state estimate in traditional search) and backtracking opportunities. For standard A* variants, the actual impact of overweighting the heuristic function towards non-admissibility varies greatly according to the stage of search progression at which it is applied and suggests that the options for intervention should take place over the early stage of the search progression, and this could be the case across a range of search problems (Figure 1b).

This is the solution we have adopted in previous work, also owing to the response time of the fNIRS signal: it could however still be of interest even when using EEG-based input frontal asymmetry scores because of the signal dynamics and the need to stabilise it over the NF epoch. Moving towards some interruptible, anytime-like approach could bring the further advantage of buffering the BCI input rather than constraining the user input in terms of timing and dynamics. A particular implementation of the above consists of parameterising heuristic search from user profiling data prior to triggering AI computation. An essential condition for this parameterisation is the availability of a framework to unify search behaviour with user personality traits that would be readily accessible through BCI measurements. One such example would make use of prefrontal asymmetry under its electrical signal form (EEG), which has been shown to have trait properties [17], to characterise, in context, user disposition towards gain, reward or, risk. On the AI side, the above user dispositions can be interpreted as potential acceptance of various forms of suboptimal solutions. These dispositions could be translated into non-admissible variants of heuristic search trading optimality for speed. One more specific case would be the explicit use of a user's risk propensity profile to be mapped onto an interpretation of risk in a heuristic search. Another core element of the system design is the timing and duration of BCI input. This design faces a number of constraints, from the user's response time in assessing the progression of the AI computation to the onset of BCI signals and any difficulty in sustaining it. In addition, difficulties in controlling the magnitude of the BCI signal may be offset by repeated interventions throughout AI computation, subject to constraints on the intervention window for offline heuristic search.

The difficulty in sustaining the BCI input signal is amply discussed in the NF literature and is one of the reasons for defining NF epochs of limited duration [59]. Moreover, even across defined NF training sessions, many recent papers have noted a drop in user BCI performance towards the latter epochs, which they have explicitly attributed to BCI fatigue. The difficulty for users to exert sustained control over specific brain regions activation is at the heart of BCI usability limitations. Leaving aside individual differences in ability, sometimes referred to as BCI illiteracy or non-responsiveness, which can be generic or specific to some BCI configurations, even the performance of a responsive subject tends to be inconsistent across trials. User task fatigue [60] has been particularly well documented during NF training involving a fixed sequence of epochs, with the performance of even good responders waning towards the last epochs of a training session. A practical consequence for BCI-controlled search would be to limit the number of user interventions in the course of any problem-solving session, as well as their duration.

2.4. Visualisation of Search Dynamics and User Response

There are a limited number of cases for which the problem being solved can be usefully visualised to give the user access to search progression towards a solution. Among the determinants making this possible are the spatial nature of the problem, the level of backtracking and monotonicity of solution construction, and the ability to derive a semantic interpretation from the search visualisation. One of the most straightforward examples is the use of heuristic search in path planning where the search progression can be visualised in real-time on the discretisation grid that supports the search process (see below, Figure 5). The overall progression can be made even more visible for complex obstacle densities and high probability of backtracking by highlighting those nodes of the grid that constitute the OPEN list. On the other hand, the tree-based visualisation of the search space of a puzzle (e.g., n-puzzle, Rubik's cube . . .) is unlikely to offer sufficient insight to the user owing to the amount of information, difficulty of interpretation, and speed of search space expansion that generally exceeds human processing abilities in the absence of high-level detectable patterns. Such patterns are similar to those which would be encountered in board games but may only be visible to experienced players: in any case, we are not dealing here with adversarial heuristic search.

It is generally accepted that the feedback element of NF-based BCI helps the user in sustaining the activation of the target region of interest, even more so that the target is not under direct volitional control. This aspect has been discussed in the NF literature from multiple perspectives: the use and type of cognitive strategies, the classification of subjects into responders and non-responders, an ability to control the BCI signal that improves during training and the number of training sessions, and the positive impact of realistic feedback channels (e.g., games, virtual reality) over abstract visual indicators [61]. The ideal, long-term configuration, would be to use the visualisation of search progression itself as the NF signal: however, a major challenge to implement this approach would be to align the temporal aspects and sampling rates of the input BCI signal and the feedback signal.

To a large extent, BCI-controlled search aims at influencing the exploration of the search space. It would then appear logical to present the users with some representation of the search space itself so that they would respond to the global shape of the search space from the initial to goal state. Assuming primarily a tree-based expansion, the traditional representation of heuristic search space is triangular (see for instance [45] (p. 152)). Moreover, the simple geometric shape and its natural interpretation in terms of 'focussing' the search to reduce the search space and expand more directly towards the goal can support a direct BCI feedback in the framework of a NF approach to BCI input, which has been shown to be appropriate to signals such as prefrontal asymmetry (Figure 4).

Although less immediately visual than the above abstract representations, search progression can also be represented through the time variation of the heuristic function values from the initial state to the solution state. It is only meaningful in terms of prompting user intervention when the heuristic shows a regular, ideally monotonic, trajectory towards the goal, such as on Figure 3b. On the other hand, heuristic value oscillations such as the one observed on Figure 3a for a classical 8-puzzle problem are not good candidates for such visualisation, because they do not converge until the very latest stages of the search.

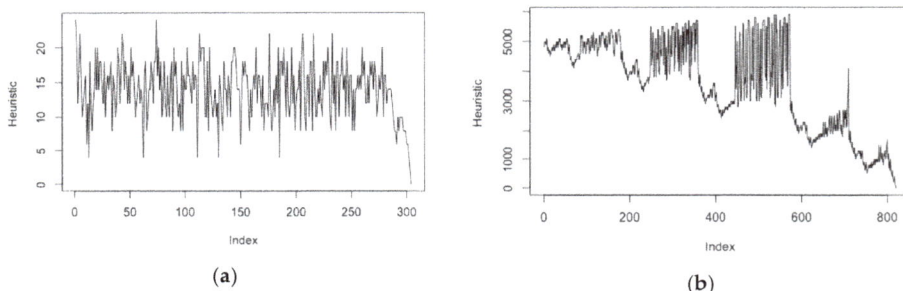

(a)   (b)

Figure 3. The variation of the heuristic function throughout the search process conditions the type of intervention (**a**) For the 8-puzzle problem, the heuristic function (Manhattan distance for misplaced tiles) oscillates significantly with search backtracking; (**b**) For a path planning problem, such as the one used in our preliminary experiments, there is an overall trend for the heuristic function (straight-line distance in arbitrary grid units) as the path progresses towards the goal node.

Influencing AI systems, as reported here, assumes a compatibility of timescales between AI computations, user perception of solution progression, and time constraints of NF input (response time, signal stability, and duration of an epoch). Despite progress made in AI techniques, typical search, planning, and optimisation problems still often require minutes of intensive computations to reach a result, as illustrated by standard benchmarks such as in the international planning competitions [62], where a cut-off time of 1800 s is introduced [63,64]. These timescales are much more representative of the target applications for our approach than examples such as the 8-puzzle used for proof-of-concept, which tend to be solvable in a few seconds. However, it should still be noted that the A* algorithm still today cannot scale up beyond simple problems [65], making non-admissible search and our overall approach still relevant.

With NF epochs generally under 60 s, we would suggest that such timescales are close to optimality when it comes to designing human intervention, in particular for those problems exhibiting a heuristic progression profile such as the one of Figure 3b (which matches that of search-based planning (e.g., in [66])).

Brain Sci. **2018**, *8*, 166

3. Proof-of-Concept Experiments: BCI Control of Heuristic Search

In order to validate our motivational model, we carried out proof-of-concept experiments in which users could influence the course of heuristic search calculations using BCI input.

The motivational framework consists of trading solution optimality for speed of calculation: on the AI side, it is implemented through ε-bound heuristic search, and on the BCI side, the motivational element is captured through real-time variations of PFC asymmetry, measured using fNIRS NF. There is ample evidence that subjects can alter prefrontal asymmetry in real-time under a NF paradigm, using various cognitive strategies [27,56,67], some of which are clearly motivational (approach-based). We have in previous work successfully used prefrontal asymmetry as a BCI paradigm using both EEG with fMRI validation [68] and fNIRS [29,55].

Overall, the system comprises the AI component, which consists of a non-admissible A* implementation in the form of weighted A* (WA*) [47] operating on a standard heuristic search problem (8-puzzle or grid-based path planning), the fNIRS-based BCI interface that measures variations of prefrontal asymmetry from a baseline under a neurofeedback paradigm, a visualisation environment that supports the NF response and gives insight into the search space of WA*, and a mapping algorithm, which determines which variations of WA* weighting coefficients should be applied for the current variation of prefrontal asymmetry.

The main objective of these experiments was to validate the motivational framework by showing that the BCI input can provide the necessary influence over the heuristic search computation in terms of information, bandwidth, and timing. Although this demonstrator does not yet implement all of the framework elements introduced in this paper (in particular, in terms of interaction timing and dynamics in relation to heuristic search progression), one important objective is to demonstrate some quantitative aspects of the mapping between BCI and heuristic search, namely that the magnitude of the user input can actually drive the computation towards various trade-offs between optimality and speed. We use one single integration paradigm, which is the precision–admissibility trade-off [45], also known as the optimality–time trade-off in cognitive research [50], where it is considered a motivational, approach-based implementation.

The common setting for the proof-of-concept experiments is based on a BCI NF paradigm, where active biofeedback is meant to support the user in controlling his/her prefrontal asymmetry. This is based on a large body of work that has demonstrated that prefrontal asymmetry could be controlled through NF across various types of BCI, electric (EEG) [68] or metabolic, in particular rt-fMRI [27]. In addition, previous research has established that the DLPFC, considered the main region involved in motivation-based PFC asymmetry [25,27], is readily accessible through fNIRS [69], including fNIRS NF [67,70]. Our NF protocol is primarily inspired by the rt-fMRI experiments on PFC asymmetry of [27], which helped us in defining epoch durations, time delays, magnitude of signal variation, and statistical validation. We have previously validated fNIRS PFC NF in a typical PFC asymmetry context dissociating approach from valence, which detected the expression of anger [29].

The NF experiments are organised around specific sessions in which NF facilitates BCI input to influence AI computation: each session is composed of various blocks that enable baseline activity definition and BCI input itself. The details of block design and experiments can be found in [71] and are only briefly described here (see also Figure 4). The generic principle consists of having a single NF block compatible with the timing of fNIRS variations and serving as BCI input to control the AI computation. fNIRS being a metabolic method, there is no absolute baseline for PFC asymmetry like the one that exists in EEG measurements, imposing to recalculate a baseline asymmetry value before each NF block. Depending on the experiment, the baseline involves rest or an unrelated cognitive task (counting) not affecting PFC asymmetry. The asymmetry score computed during the baseline is used as reference and considered "zero asymmetry" regardless of its actual value. The last 10 s of the resting epoch (Figure 4) are used to measure that score with specific care taken not to induce variations of asymmetry.

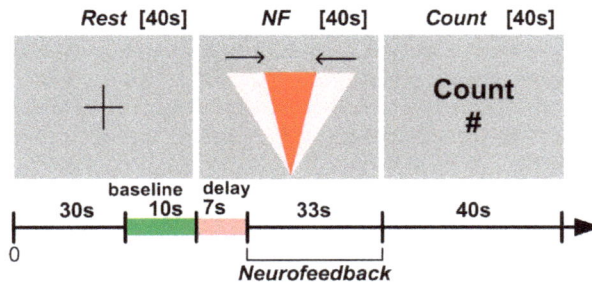

Figure 4. The neurofeedback (NF) protocol used for the 8-puzzle experiment. Note that the last 10s of a resting epoch are used to determine the prefrontal cortex (PFC) asymmetry baseline a priori to the NF epoch. The 7-s delay is introduced to take into account the onset of haemodynamic response in fNIRS. The NF epoch is followed by a non-motivational cognitive task facilitating the return to a new baseline [71].

As with all NF installations, the feedback signal should be determined by the level of activation of the region of interest (here, the difference in activation between left and right PFC calculated by averaging oxy-haemoglobin (HbO) values over the four leftmost and four rightmost fNIRS channels, then subtracting the average right from the average left).

The first experiment explored BCI control over heuristic search for solutions to the 8-puzzle (Figure 5). The rationale for using a textbook example such as the 8-puzzle is that its complete solution set is fully accessible [54], which considerably simplifies the experimental design by selecting 8-puzzle configurations (starting state and goal state) whose properties are known.

For instance, when applying heuristic weighting modifications during the search itself, it is possible to experiment with known solution lengths or configurations, admitting a large number of solutions to minimise the impact of dynamic modifications. Because the range of solutions and impact of ε-admissibility is documented, the mapping of BCI input to heuristic search is also easier to describe and experiment with.

The mechanism by which a feedback signal is generated from the detection of BCI input is generally referred to as mapping and plays an important role in NF design (Figure 5(4)). Here, the starting point to determine the best mapping functions is to look at the outcome of non-admissible search experiments. These determine the range of heuristic function modifications that have the most significant effect in terms of performance–admissibility trade-off. Previous literature on non-admissible search [47,48] has established a number of principles, such as the fact that the main impact of non-admissibility is to reduce the size of the search space or that significant effects could be observed for even minor modifications of the heuristic weighting. In our experiments, the BCI signal (level of asymmetry compared to the baseline) is mapped linearly onto an abstract symbology for the search space taking the form of a two-dimensional (2D) beam whose width represents the variable to be minimised. We have based the mapping on the statistical significance of fNIRS signal variation with respect to the baseline using real-time t-tests and associated effect size (Figure 5(3)). The post-hoc validation of each NF epoch has been confirmed using resampling methods, in particular bootstrapping [72].

The intervention model for the 8-puzzle was to request a NF intervention soon after the start of the search process, resulting in the heuristic weighting being altered after 0–25% of the search space had been explored (this value being derived from the known solution configuration, see Figure 1).

In the case of the 8-puzzle, the main impact was on search space reduction, measured through a reduction in the number of nodes expanded [48] and consistent with our preliminary tests of non-admissible search (Figure 5(6)). It is worth noting that the optimality of the solution was actually often preserved, meaning that the users were actually successful in speeding up the AI computation without compromising solution quality.

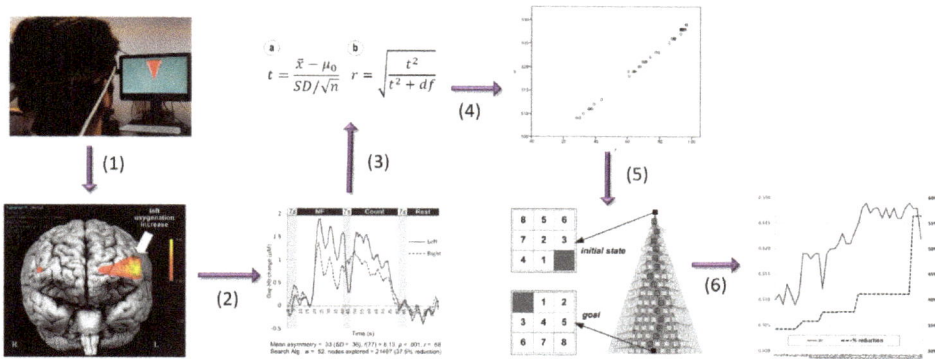

$$t = \frac{\bar{x} - \mu_0}{SD/\sqrt{n}} \quad r = \sqrt{\frac{t^2}{t^2 + df}}$$

Figure 5. Brain–computer interfaces (BCI) interfacing to heuristic search on the 8-puzzle. The user's motivational dimension is obtained through fNIRS measurement of PFC asymmetry (**1**). The level of change from the baseline, which is taken to measure approach, is determined with real-time statistical testing (**2,3**). It is mapped linearly onto the WA* weighting parameter using the effect size to determine the level of heuristic modification (**4**). The change in weighting parameter for WA* is applied during the search (**5**), which results in search space reduction and computation speed-up (**6**). Note the abstract representation of the search space as a two-dimensional (2D) beam (**1**), which serves as a visual feedback for fNIRS NF (adapted from [71]).

The variations in prefrontal asymmetry across subjects resulted in differentiated effects on heuristic weighting and associated search space reduction, compatible with the intended quantified use. However, there was not enough data in our single-trial experiments to assess intrasubject variations and validate how a single subject could fine-tune the behaviour of a given search progress. This raises the issue of the controllability of the magnitude effect, which should be the object of further experiments but could also be mitigated through multiple interventions during a given AI computation.

A second set of experiments was staged using grid-based path planning as a heuristic search problem (Figure 6). The rationale for this second test case was that the search space could be visualised in real time as the search progressed so that the visual feedback sent to the user about search progression was no longer metaphorical. However, to avoid potential uncanny effects due to the shape of the node frontiers progression (which with grid-based path planning also depends on obstacle density and environment layout), the display superimposed the same triangular shape over the node progression to be used as the NF channel. In this second experiment, the search space is comparatively smaller, and the reduction in search space is less dramatic than with the 8-puzzle. However, non-admissible search produces qualitative, as well as quantitative modifications of the solution path, which can be readily observed on the chosen obstacle configuration: the solution path under user intervention is more straightforward and travels through the centre of the environment.

Interestingly, the success rates did not differ significantly from the 8-puzzle experiments, suggesting that a better visibility of the search space progression did not improve subjects' performance in that instance. However, this might depend on the actual obstacle density and layout, as the actual shape of the front node progression and associated backtracking might actually be distracting to users.

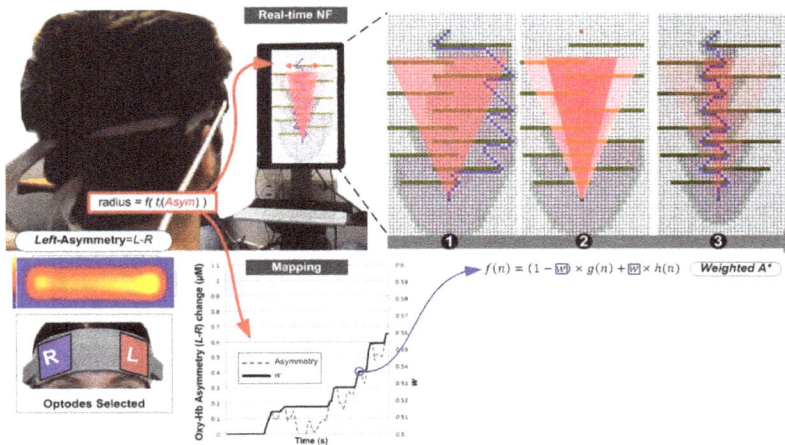

Figure 6. BCI interfacing to heuristic search on a path planning problem. The motivational dimension is acquired through fNIRS-based prefrontal cortex asymmetry. In this experiment, the heuristic function can be repeatedly modified as the search progresses, also taking advantage of the more visual feedback provided by path progression. Because of the multiple updates, in this experiment the weighting factor has only been allowed to increase through time to explore search speed-up. Note the change in the qualitative nature of the solution (path geometry) from solution (1) to solution in (3). From [71].

The users' perception of the task can be analysed through their narrative feedback on the cognitive strategies they used to increase prefrontal asymmetry. Several users reported strategies compatible with approach and result anticipation such as imagining running in a virtual race or encouraging the progression of the search as one would encourage a racer. Prior to the experiment, subjects were explained the goal of AI computation and the NF setting, although we refrained from suggesting explicit cognitive strategies. However, a few others mentioned the recollection of positive autobiographic memories, which is known to also induce left prefrontal asymmetry because of the interplay between valence and approach in appetitive stimuli or recollections, as also reported by Zotev et al. [27] in their fMRI prefrontal asymmetry NF experiments.

In these experiments, NF success is defined for each subject as having at least half of successful blocks during a NF trial [72]: this high-level measure is meant to give an indication on the usability of the interactive system.

It is interesting to compare current success scores to two other previous fNIRS experiments also involving PFC asymmetry in two different affective contexts (engagement (Aranyi et al. [55]) and anger (Aranyi et al. [29])). All these experiments have in common a minimal level of user training which tends to be the same across experiments: the calculation of PFC asymmetry from haemodynamic data is similar, based on the same optodes and the same formula. Previous affective BCI experiments resulted in success scores of 73% [55] and 70% [29]. Our new 8-puzzle experiment achieved a similar score of 73%, suggesting that significant NF success is possible in the absence of a clear affective context, with a motivational-based approach for which there is no priming from the application or visual environment. Paradoxically, the increased visual realism in the path planning setting did not result in higher success scores, despite the reported positive impact of visual realism on NF [61]. Based on debriefing and narrative feedback from the subjects, the lower success scores observed for path planning (57%) were attributable to the extra cognitive load induced by the visual complexity. Another potential explanation is that in the path planning experiment, the baseline was determined during the counting epoch rather than during a post-counting resting epoch; although counting is considered a neutral task for prefrontal asymmetry, it could in some cases affect it via mental workload for some subjects [73], thereby introducing a ceiling effect in PFC asymmetry variation with subsequent impact on success scores.

Another important point to consider when analysing performance is that we are using NF as an interaction paradigm rather than as a therapeutic approach. Of NF, we only retain the hypothesis according to which the presence of the feedback signal helps the user activate brain regions not directly accessible to volitional control. Unlike NF therapeutic systems we do not include multiple training sessions, which are used to induce long-term behavioural changes (mediated by neural plasticity) and are generally associated with an improvement in the ability to control the NF signal throughout training. This induces an inherent limitation in our approach, which is that overall subject performance will generally be lower in the absence of multiple training sessions. The minimal training provided to our subjects can be counted in minutes, whilst it is generally considered that several hours (up to 40, [74]) through repeated sessions are required for subjects to be confident with NF control. In practice, subjects were allowed between one (path planning) and three (8-puzzle) blocks for training, which, considering the maximum block length of 120 s, can safely be considered as mere familiarisation rather than training across multiple sessions.

One objective of the proof-of-concept experiments was to demonstrate the users' ability to control prefrontal asymmetry in a generic motivational context related to the expectation of a computation result, this expectation taking the shape of a trade-off between quality and performance. This objective is highly specific to the possibility to control AI systems and differs from previous BCI use of PFC asymmetry, which has been primarily involved with affective BCI [29,55]. This difference arises from the generic motivational model associated with PFC asymmetry, which can be connected both to reward expectation and to appetitive stimuli, the latter going as far on the affective spectrum as to constitute a high-level dimensional aspect for empathy. In all our previous affective BCI work, a strong context, both prior to the NF trials and during trials themselves, may have facilitated user control. For instance, in eliciting anger against a virtual agent, subjects have been shown short videos evidencing the bad character of the agent [29]; in eliciting empathy or support, they have followed a narrative showing the character in trouble [68]. No such context is available when considering the control of algorithmic AI progression: moreover, as we are using abstract benchmark examples that do not even correspond to popular board games, it appears essential to assess how users can operate in the absence of a direct sense of reward expectation, other than the one conveyed to them as part of the experiment brief.

4. Conclusions and Further Work

We have introduced a framework inspired by human augmentation for the control of autonomous AI systems, which opens the way to the development of new interaction technologies dedicated to human–AI cooperation. This framework departs from previous research in that it seeks to adapt to the imbalance between high-performance autonomous AI systems and users' information processing abilities and response times, which require the latter to operate at specific levels of abstraction. The description of this framework has uncovered a number of important design issues, amongst which are the synchronisation of BCI input and AI computations and the leverage effect that basic AI mechanisms such as search will have on global computation. The former aspect will prescribe under which conditions BCI-input delays provided by metabolic methods such as fNIRS can be accommodated or whether the system should resort to EEG measurements of motivational parameters. Our proof-of-concept experiments have only examined traditional search problems, without addressing the potential leverage that the search will bring onto higher-level AI computations. One candidate technique to further this aspect of the research would be to examine heuristic search planning systems [43]. One notable element is that the heuristic function in some heuristic search planning applications tends to follow a trend similar to that of Figure 3b [66].

Throughout our early work, we have opted for single NF sessions with a limited number of epochs, supported by cognitive strategies. Although we have not been prescriptive about the type

Brain Sci. **2018**, *8*, 166

of cognitive strategies to be used, we have introduced subjects to the concept of cognitive strategy, as "thought contents" that would lead to best performance in the NF task. The role played by cognitive strategies can be explained in part by the fact that these experiments implemented single-session NF: although the actual requirement for cognitive strategies in NF has been debated [75], repeated training sessions may be required for subjects to perform without the help of a cognitive strategy.

It now appears that too much emphasis on cognitive strategies may actually distract users from the observation of AI computation progress, which should be the primary driver of their BCI input. In the future, this could be addressed through two complementary directions. One would consist of a more comprehensive use of the AI computation progress as a visualisation feedback channel to support BCI input: however, this approach would require non-trivial temporal alignment between AI progression visualisation and the NF interface, which could require buffering, warping, or predictive features to be incorporated. Another direction is to accept the need for extensive NF training to support users' performance: typical training times reported range from a few hours to up to 40 h [74].

Even restricting ourselves to a motivational model, it is not always possible to distinguish whether variations in prefrontal asymmetry should be interpreted in terms of approach [17] or in terms of risk taking [52]. This is part of a broader issue, well described in prefrontal asymmetry research, known as the balance of activity variation across each hemisphere that accounts for the observed increase in left asymmetry (because left asymmetry is the target in our experiments). During our previous experiments on PFC asymmetry [29,55,68], most of the increase in prefrontal asymmetry could be attributed to a proportionally greater increase in left-side rather than right-side activity. It has proven elusive to observe a selective decrease of right PFC activity, even a relative one, as a mechanism for left asymmetry, including in the experiments upon which we are commenting here, suggesting that increased risk taking cannot be considered as a primary mechanism. However, recent EEG NF work has evidenced such selective decrease in right prefrontal activity [57]. If this latter effect could be reproduced in a hybrid cognitive scenario, it could open the way to a risk–acceptance paradigm, as discussed above. A successful implementation of a risk paradigm would have significant interest in terms of AI applications, provided it ensures that users have an appropriate perception of alternative solutions in terms of risks.

Acknowledgments: Gabor Aranyi and Fred Charles had an essential contribution to the proof-of-concept experiments whose results are discussed in Section 3 of this paper: details of their respective contributions are listed in reference [71].

Conflicts of Interest: The author declares no conflict of interest.

References

1. Saproo, S.; Faller, J.; Shih, V.; Sajda, P.; Waytowich, N.R.; Bohannon, A.; Lawhern, V.J.; Lance, B.J.; Jangraw, D. Cortically coupled computing: A new paradigm for synergistic human-machine interaction. *Computer* **2016**, *49*, 60–68. [CrossRef]

2. Eugster, M.J.A.; Ruotsalo, T.; Spapé, M.M.; Kosunen, I.; Barral, O.; Ravaja, N.; Jacucci, G.; Kaski, S. Predicting term-relevance from brain signals. In Proceedings of the 37th International ACM SIGIR Conference on Research & Development in Information Retrieval, Gold Coast, Australia, 6–11 July 2014; pp. 425–434.

3. Gerson, A.D.; Parra, L.C.; Sajda, P. Cortically coupled computer vision for rapid image search. *IEEE Trans. Neural Syst. Rehabil. Eng.* **2006**, *14*, 174–179. [CrossRef] [PubMed]

4. Ferrucci, D.; Levas, A.; Bagchi, S.; Gondek, D.; Mueller, E.T. Watson: Beyond jeopardy! *Artif. Intell.* **2013**, *199*, 93–105. [CrossRef]

5. Silver, D.; Huang, A.; Maddison, C.J.; Guez, A.; Sifre, L.; van den Driessche, G.; Schrittwieser, J.; Antonoglou, L.; Panneershelvam, V.; Lanctot, M.; et al. Mastering the game of Go with deep neural networks and tree search. *Nature* **2016**, *529*, 484–489. [CrossRef] [PubMed]

6. Brynjolfsson, E.; McAfee, A. *The Second Machine Age: Work, Progress, and Prosperity in a Time of Brilliant Technologies*; WW Norton & Company: New York, NY, USA, 2014.

7. Sharkey, N. Cassandra or false prophet of doom: AI robots and war. *IEEE Intell. Syst.* **2008**, *23*, 14–17. [CrossRef]

8. Boström, N.; Yudkowsky, E. The ethics of artificial intelligence. *Camb. Handb. Artif. Intell.* **2014**, *316*, 334.

9. Boström, N. *Superintelligence: Paths, Dangers, Strategies*; Oxford University Press: Oxford, UK, 2014.

10. Kennedy, P. Brain-machine interfaces as a challenge to the "moment of singularity". *Front. Syst. Neurosci.* **2014**, *8*, 213. [CrossRef] [PubMed]

11. Skulimowski, A.M.J. Future prospects of human interaction with artificial autonomous systems. In Proceedings of the International Conference on Adaptive and Intelligent Systems, Bournemouth, UK, 8–9 September 2014; pp. 131–141.

12. Barrett, A.M.; Baum, S.D. A model of pathways to artificial superintelligence catastrophe for risk and decision analysis. *J. Exp. Theor. Artif. Intell.* **2017**, *29*, 397–414. [CrossRef]

13. Jacucci, G.; Spagnolli, A.; Freeman, J.; Gamberini, L. Symbiotic interaction: A critical definition and comparison to other human-computer paradigms. In Proceedings of the International Workshop on Symbiotic Interaction, Eindhoven, The Netherlands, 18–19 December 2014; pp. 3–20.

14. Gaggioli, A.; Ferscha, A.; Riva, G.; Dunne, S.; Viaud-Delmon, I. *Human Computer Confluence: Transforming Human Experience through Symbiotic Technologies*; De Gruyter Open: Berlin, Germany, 2016.

15. Harmon-Jones, E.; van Honk, J. Introduction to a special issue on the neuroscience of motivation and emotion. *Motiv. Emot.* **2012**, *36*, 1–3. [CrossRef]

16. Braver, T.S.; Krug, M.K.; Chiew, K.S.; Kool, W.; Westbrook, J.A.; Clement, N.J.; Adcock, R.A.; Barch, D.M.; Botvinick, M.M.; Carver, C.S.; et al. Mechanisms of motivation–cognition interaction: Challenges and opportunities. *Cogn. Affect. Behav. Neurosci.* **2014**, *14*, 443–472. [CrossRef] [PubMed]

17. Harmon-Jones, E.; Gable, P.A. On the role of asymmetric frontal cortical activity in approach and withdrawal motivation: An updated review of the evidence. *Psychophysiology* **2018**, *55*, e12879. [CrossRef] [PubMed]

18. Gollwitzer, P.M. Mindset Theory of Action Phases. In *Handbook of Theories of Social Psychology*; van Lange, P., Kruglanski, A.W., Higgins, E.T., Eds.; Sage: London, UK, 2012; Volume 1, pp. 526–545.

19. Harmon-Jones, E.; Gable, P.A.; Peterson, C.K. The role of asymmetric frontal cortical activity in emotion-related phenomena: A review and update. *Biol. Psychol.* **2010**, *84*, 451–462. [CrossRef] [PubMed]

20. Sutton, S.K.; Davidson, R.J. Prefrontal brain asymmetry: A biological substrate of the behavioral approach and inhibition systems. *Psychol. Sci.* **1997**, *8*, 204–210. [CrossRef]

21. Smith, E.E.; Reznik, S.J.; Stewart, J.L.; Allen, J.J.B. Assessing and conceptualizing frontal EEG asymmetry: An updated primer on recording, processing, analyzing, and interpreting frontal alpha asymmetry. *Int. J. Psychophysiol.* **2017**, *111*, 98–114. [CrossRef] [PubMed]

22. Aupperle, R.L.; Melrose, A.J.; Francisco, A.; Paulus, M.P.; Stein, M.B. Neural substrates of approach-avoidance conflict decision-making. *Hum. Brain Mapp.* **2015**, *36*, 449–462. [CrossRef] [PubMed]

23. Gorka, S.M.; Phan, K.L.; Shankman, S.A. Convergence of EEG and fMRI measures of reward anticipation. *Biol. Psychol.* **2015**, *112*, 12–19. [CrossRef] [PubMed]

24. Amodio, D.M.; Shah, J.Y.; Sigelman, J.; Brazy, P.C.; Harmon-Jones, E. Implicit regulatory focus associated with asymmetrical frontal cortical activity. *J. Exp. Soc. Psychol.* **2004**, *40*, 225–232. [CrossRef]

25. Berkman, E.T.; Lieberman, M.D. Approaching the bad and avoiding the good: Lateral prefrontal cortical asymmetry distinguishes between action and valence. *J. Cogn. Neurosci.* **2010**, *22*, 1970–1979. [CrossRef] [PubMed]

26. Sherwood, M.S.; Kane, J.H.; Weisend, M.P.; Parker, J.G. Enhanced control of dorsolateral prefrontal cortex neurophysiology with real-time functional magnetic resonance imaging (rt-fMRI) Neurofeedback training and working memory practice. *Neuroimage* **2016**, *124*, 214–223. [CrossRef] [PubMed]

27. Zotev, V.; Phillips, R.; Yuan, H.; Misaki, M.; Bodurka, J. Self regulation of human brain activity using simultaneous real-time fMRI and EEG NF. *NeuroImage* **2014**, *85*, 985–995. [CrossRef] [PubMed]

28. Doi, H.; Nishitani, S.; Shinohara, K. NIRS as a tool for assaying emotional function in the prefrontal cortex. *Front. Hum. Neurosci.* **2013**, *7*, 770. [CrossRef] [PubMed]

29. Aranyi, G.; Charles, F.; Cavazza, M. Anger-based BCI using fNIRS NF. In Proceedings of the 28th Annual ACM Symposium on User Interface Software & Technology, Charlotte, NC, USA, 11–15 November 2015; pp. 511–521.

30. Kelley, N.J.; Hortensius, R.; Schutter, D.J.L.G.; Harmon-Jones, E. The relationship of approach/avoidance motivation and asymmetric frontal cortical activity: A review of studies manipulating frontal asymmetry. *Int. J. Psychophysiol.* **2017**, *119*, 19–30. [CrossRef] [PubMed]

31. Fecteau, S.; Knoch, D.; Fregni, F.; Sultani, N.; Boggio, P.; Pascual-Leone, A. Diminishing risk-taking behavior by modulating activity in the prefrontal cortex: A direct current stimulation study. *J. Neurosci.* **2007**, *27*, 12500–12505. [CrossRef] [PubMed]

32. Nejati, V.; Salehinejad, M.A.; Nitsche, M.A. Interaction of the left dorsolateral prefrontal cortex (l-DLPFC) and right orbitofrontal cortex (OFC) in hot and cold executive functions: Evidence from transcranial direct current stimulation (tDCS). *Neuroscience* **2018**, *369*, 109–123. [CrossRef] [PubMed]

33. Pizzagalli, D.A.; Sherwood, R.J.; Henriques, J.B.; Davidson, R.J. Frontal brain asymmetry and reward responsiveness: A source-localization study. *Psychol. Sci.* **2005**, *16*, 805–813. [CrossRef] [PubMed]

34. Rolls, E.T.; Grabenhorst, F. The orbitofrontal cortex and beyond: From affect to decision-making. *Prog. Neurobiol.* **2008**, *86*, 216–244. [CrossRef] [PubMed]

35. Lee, D.; Seo, H. Mechanisms of reinforcement learning and decision making in the primate dorsolateral prefrontal cortex. *Ann. N. Y. Acad. Sci.* **2007**, *1104*, 108–122. [CrossRef] [PubMed]

36. Tanaka, S.C.; Samejima, K.; Okada, G.; Ueda, K.; Okamoto, Y.; Yamawaki, S.; Doya, K. Brain mechanism of reward prediction under predictable and unpredictable environmental dynamics. *Neural Netw.* **2006**, *19*, 1233–1241. [CrossRef] [PubMed]

37. Wallis, J.D.; Miller, E.K. Neuronal activity in primate dorsolateral and orbital prefrontal cortex during performance of a reward preference task. *Eur. J. Neurosci.* **2003**, *18*, 2069–2081. [CrossRef] [PubMed]

38. Li, J.; Delgado, M.R.; Phelps, E.A. How instructed knowledge modulates the neural systems of reward learning. *Proc. Natl. Acad. Sci. USA* **2011**, *108*, 55–60. [CrossRef] [PubMed]

39. Coutlee, C.G.; Huettel, S.A. The functional neuroanatomy of decision making: Prefrontal control of thought and action. *Brain Res.* **2012**, *1428*, 3–12. [CrossRef] [PubMed]

40. Korf, R.E. Finding optimal solutions to Rubik's cube using pattern databases. In Proceedings of the Fourteenth National Conference on Artificial Intelligence and Ninth Conference on Innovative Applications of Artificial Intelligence, Providence, RI, USA, 27–31 July 1997; pp. 700–705.

41. Zeng, Y.; Chen, X.; Cao, X.; Qin, S.; Cavazza, M.; Xiang, Y. Optimal Route Search with the Coverage of Users' Preferences. In Proceedings of the Twenty-Fourth International Joint Conference on Artificial Intelligence, Buenos Aires, Argentina, 25–31 July 2015; pp. 2118–2124.

42. Stern, R.; Lelis, L.H.S. What's Hot in Heuristic Search. In Proceedings of the Thirtieth AAAI Conference on Artificial Intelligence, Phoenix, AZ, USA, 12–17 February 2016; pp. 4340–4342.

43. Bonet, B.; Geffner, H. Planning as heuristic search. *Artif. Intell.* **2001**, *129*, 5–33. [CrossRef]

44. Fader, A.; Zettlemoyer, L.; Etzioni, O. Open question answering over curated and extracted knowledge bases. In Proceedings of the 20th ACM SIGKDD International Conference on Knowledge Discovery and Data Mining, New York, NY, USA, 24–27 August 2014; pp. 1156–1165.

45. Pearl, J. *Heuristics: Intelligent Search Strategies for Computer Problem Solving*; Addison-Wesley: Reading, MA, USA, 1984.

46. Pohl, I. The avoidance of (relative) catastrophe, heuristic competence, genuine dynamic weighting and computational issues in heuristic problem solving. In Proceedings of the 3rd International Joint Conference on Artificial Intelligence, Stanford, CA, USA, 20–23 August 1973; pp. 12–17.

47. Ebendt, R.; Drechsler, R. Weighted A* search–unifying view and application. *Artif. Intell.* **2009**, *173*, 1310–1342. [CrossRef]

48. Hansen, E.A.; Zhou, R. Anytime heuristic search. *J. Artif. Intell. Res.* **2007**, *28*, 267–297. [CrossRef]

49. Barer, M.; Sharon, G.; Stern, R.; Felner, A. Suboptimal variants of the conflict-based search algorithm for the multi-agent pathfinding problem. In Proceedings of the Seventh Annual Symposium on Combinatorial Search, Praha, Czech Republic, 15–17 August 2014.

50. Bijleveld, E.; Custers, R.; Aarts, H. Unconscious reward cues increase invested effort, but do not change speed–accuracy tradeoffs. *Cognition* **2010**, *115*, 330–335. [CrossRef] [PubMed]

51. Wilt, C.M.; Ruml, W. When Does Weighted A* Fail? In Proceedings of the Fifth Annual Symposium on Combinatorial Search, Niagara Falls, ON, Canada, 19–21 July 2012; pp. 137–144.

52. Santesso, D.L.; Segalowitz, S.J.; Ashbaugh, A.R.; Antony, M.M.; McCabe, R.E.; Schmidt, L.A. Frontal EEG asymmetry and sensation seeking in young adults. *Biol. Psychol.* **2008**, *78*, 164–172. [CrossRef] [PubMed]

53. Likhachev, M.; Gordon, G.J.; Thrun, S. ARA*: Anytime A* with provable bounds on sub-optimality. *Adv. Neural Inf. Process. Syst.* **2004**, *16*, 767–774.

Brain Sci. **2018**, *8*, 166

54. Reinefeld, A. Complete Solution of the Eight-Puzzle and the Bene t of Node Ordering in IDA*. In Proceedings of the 3th International Joint Conference on Artifical Intelligence, Chambery, France, 28 August–3 September 1993; pp. 248–253.

55. Aranyi, G.; Pecune, F.; Charles, F.; Pelachaud, C.; Cavazza, M. Affective interaction with a virtual character through an fNIRS brain-computer interface. *Front. Comput. Neurosci.* **2016**, *10*, 70. [CrossRef] [PubMed]

56. Rosenfeld, J.P.; Cha, G.; Blair, T.; Gotlib, I.H. Operant (biofeedback) control of left-right frontal alpha power differences: Potential neurotherapy for affective disorders. *Biofeedback Self-Regul.* **1995**, *20*, 241–258. [CrossRef] [PubMed]

57. Mennella, R.; Patron, E.; Palomba, D. Frontal alpha asymmetry NF for the reduction of negative affect and anxiety. *Behav. Res. Ther.* **2017**, *92*, 32–40. [CrossRef] [PubMed]

58. Korf, R.E. Real-time heuristic search. *Artif. Intell.* **1990**, *42*, 189–211. [CrossRef]

59. Rogala, J.; Jurewicz, K.; Paluch, K.; Kublik, E.; Cetnarski, R.; Wróbel, A. The Do's and Don'ts of NF training: A review of the controlled studies using healthy adults. *Front. Hum. Neurosci.* **2016**, *10*, 301. [CrossRef] [PubMed]

60. Ekandem, J.I.; Davis, T.A.; Alvarez, I.; James, M.T.; Gilbert, J.E. Evaluating the ergonomics of BCI devices for research and experimentation. *Ergonomics* **2012**, *55*, 592–598. [CrossRef] [PubMed]

61. Cohen, A.; Keynan, J.N.; Jackont, G.; Green, N.; Rashap, I.; Shani, O.; Charles, F.; Cavazza, M.; Hendler, T.; Raz, G. Multi-modal virtual scenario enhances NF learning. *Front. Robot. AI* **2016**, *3*, 52. [CrossRef]

62. Vallati, M.; Chrpa, L.; Mccluskey, T.L. What you always wanted to know about the deterministic part of the International Planning Competition (IPC) 2014 (but were too afraid to ask). *Knowl. Eng. Rev.* **2018**, *33*. [CrossRef]

63. Percassi, F.; Gerevini, A.E.; Geffner, H. Improving Plan Quality through Heuristics for Guiding and Pruning the Search: A Study Using LAMA. In Proceedings of the 10th International Symposium on Combinatorial Search, Pittsburgh, PA, USA, 16–17 June 2017; pp. 144–148.

64. Rizzini, M.; Fawcett, C.; Vallati, M.; Gerevini, A.E.; Hoos, H.H. Static and dynamic portfolio methods for optimal planning: An empirical analysis. *Int. J. Artif. Intell. Tools* **2017**, *26*. [CrossRef]

65. Hatem, M.; Burns, E.; Ruml, W. Solving Large Problems with Heuristic Search: General-Purpose Parallel External-Memory Search. *J. Artif. Intell. Res.* **2018**, *62*, 233–268. [CrossRef]

66. Pizzi, D.; Lugrin, J.-L.; Whittaker, A.; Cavazza, M. Automatic generation of game level solutions as storyboards. *IEEE Trans. Comput. Intell AI Games* **2010**, *2*, 149–161. [CrossRef]

67. Sakatani, K.; Takemoto, N.; Tsujii, T.; Yanagisawa, K.; Tsunashima, H. NIRS-Based NF Learning Systems for Controlling Activity of the Prefrontal Cortex. In *Oxygen Transport to Tissue XXXV*; Van Huffel, S., Naulaers, G., Caicedo, A., Bruley, D.F., Harrison, D.K., Eds.; Springer: New York, NY, USA, 2013; pp. 449–454.

68. Gilroy, S.W.; Porteous, J.; Charles, F.; Cavazza, M.; Soreq, E.; Raz, G.; Ikar, L.; Or-Borichov, A.; Ben-Arie, U.; Klovatch, H.; et al. A Brain-Computer Interface to a Plan-Based Narrative. In Proceedings of the Twenty-third International Joint Conference on Artificial Intelligence, Beijing, China, 3–9 August 2013; pp. 1997–2005.

69. Ayaz, H.; Shewokis, P.A.; Bunce, S.; Onaral, B. An optical brain computer interface for environmental control. In Proceedings of the 2011 Annual International Conference of the IEEE Engineering in Medicine and Biology Society, Boston, MA, USA, 30 August–3 September 2011; pp. 6327–6330.

70. Sitaram, R.; Caria, A.; Birbaumer, N. Hemodynamic brain–computer interfaces for communication and rehabilitation. *Neural Netw.* **2009**, *22*, 1320–1328. [CrossRef] [PubMed]

71. Cavazza, M.; Aranyi, G.; Charles, F. BCI Control of Heuristic Search Algorithms. *Front. Neuroinf.* **2017**, *11*, 6. [CrossRef] [PubMed]

72. Aranyi, G.; Cavazza, M.; Charles, F. Using fNIRS for prefrontal-asymmetry neurofeedback: Methods and challenges. In Proceedings of the Fourth International Workshop on Symbiotic Interaction, Berlin, Germany, 7–8 October 2015; pp. 7–20.

73. Manoach, D.S.; Schlaug, G.; Siewert, B.; Darby, D.G.; Bly, B.M.; Benfield, A.; Edelman, R.R.; Warach, S. Prefrontal cortex fMRI signal changes are correlated with working memory load. *Neuroreport* **1997**, *8*, 545–549. [CrossRef] [PubMed]

74. Kotchoubey, B.; Kübler, A.; Strehl, U.; Flor, H.; Birbaumer, N. Can humans perceive their brain states? *Conscious Cogn.* **2002**, *11*, 98–113. [CrossRef] [PubMed]

75. Kober, S.E.; Witte, M.; Ninaus, M.; Neuper, C.; Wood, G. Learning to modulate one's own brain activity: The effect of spontaneous mental strategies. *Front. Hum. Neurosci.* **2013**, *7*, 695. [CrossRef] [PubMed]

brain
sciences

MDPI

Article

Prediction of Human Performance Using Electroencephalography under Different Indoor Room Temperatures

Tapsya Nayak, Tinghe Zhang [1], Zijing Mao [1], Xiaojing Xu [2], Lin Zhang [3], Daniel J. Pack [4], Bing Dong [5] and Yufei Huang [1,*]

[1] Department of Electrical and Computer Engineering, University of Texas at San Antonio, San Antonio, TX 78249, USA; ani254@my.utsa.edu (T.N.); prh169@my.utsa.edu (T.Z.); mzj168@hotmail.com (Z.M.)
[2] NSF-DOE CURRENT Center, University of Tennessee, Knoxville, TN 37996, USA; xiaojing.hsu@gmail.com
[3] SIEE, China University of Mining and Technology, Xuzhou 221116, China; cnnangua@hotmail.com
[4] College of Engineering & Computer Science, University of Tennessee, Chattanooga, TN 37403, USA; daniel-pack@utc.edu
[5] Department of Mechanical Engineering, University of Texas at San Antonio, San Antonio, TX 78249, USA; bing.dong@utsa.edu
* Correspondence: yufei.huang@utsa.edu; Tel.: +1-210-450-7260

Received: 13 April 2018; Accepted: 19 April 2018; Published: 23 April 2018

Abstract: Varying indoor environmental conditions is known to affect office worker's performance; wherein past research studies have reported the effects of unfavorable indoor temperature and air quality causing sick building syndrome (SBS) among office workers. Thus, investigating factors that can predict performance in changing indoor environments have become a highly important research topic bearing significant impact in our society. While past research studies have attempted to determine predictors for performance, they do not provide satisfactory prediction ability. Therefore, in this preliminary study, we attempt to predict performance during office-work tasks triggered by different indoor room temperatures (22.2 °C and 30 °C) from human brain signals recorded using electroencephalography (EEG). Seven participants were recruited, from whom EEG, skin temperature, heart rate and thermal survey questionnaires were collected. Regression analyses were carried out to investigate the effectiveness of using EEG power spectral densities (PSD) as predictors of performance. Our results indicate EEG PSDs as predictors provide the highest R^2 (> 0.70), that is 17 times higher than using other physiological signals as predictors and is more robust. Finally, the paper provides insight on the selected predictors based on brain activity patterns for low- and high-performance levels under different indoor-temperatures.

Keywords: human performance; performance prediction; indoor room temperature; office-work tasks; electroencephalography (EEG)

---

1. Introduction

As U.S. citizens spend more than 90% of their time indoors, indoor thermal condition is a key factor that impacts human productivity in the office [1–5]. Indoor environments and building characteristics have been reported to impact occurrences of respiratory diseases, allergy and asthma symptoms, sick building symptoms and office-work performance. It is estimated that improving the indoor environment in U.S. office buildings would result in a 0.5 to 5% increase in productivity, worth $12–$125 billion annually [6]. Thus, understanding how indoor environments affect human performance, health and emotion and developing methods to predict human performance/health in changing indoor environments have become highly important research topics that bear significant economic and sociological impact.

As our indoor daily work becomes increasingly mentally challenging, a significant aspect of the thermal-driven performance is an individual's cognitive performance, that is, the ability of an individual to effectively comprehend and perform independent decisions during complex tasks and events. Various field and laboratory studies have been conducted to investigate performance levels and changes under different thermal conditions. A study investigated in Reference [7] showed an 8% fall in sewing work productivity as indoor temperature was increased from 23.9 °C to 32.2 °C. A similar trend was observed in a case study by References [8,9] investigating the performance of employees in telecommunication offices (call center) and a reported decline in work performance by 5–7% at higher indoor temperatures; work performance was evaluated by assessing average time per call or average handling time. Similar studies were conducted to evaluate the performance of school children in References [10,11]. In the former research study, students who reported changes in thermal sensation scores from warm to neutral, performance of numerical and language task improved significantly, while the latter concluded thermal stress produces mental arousal effects thereby improving performance. In addition to these papers that studied the influence of indoor environment on office work performance, researchers have investigated physiological mechanisms and whether these mechanisms have consequences for human performance. At high temperatures, authors in Reference [12] reported that the concentration of carbon-dioxide (CO_2), by measuring end-tidal partial CO_2, is directly proportional to the increase in room temperature, which they hypothesize is the result of increased metabolism by humans in turn leading to decreased air quality. Furthermore, they observed a reduction in arterial blood oxygen saturation (SPO_2), increasing sick building syndrome (SBS) symptoms thereby elevating fatigue levels in participants. A brain imaging near-infrared spectroscopy (NIRS) study by the authors in Reference [13] observed a reduction in task performance as blood oxygen saturation levels decrease. Interestingly, while Reference [14] found decreased concentrations of salivary alpha-amylase and cortisol with increased thermal discomfort—implying an impact on performance—but performance did not change. On the other hand, they found carbon dioxide concentrations to be similar at different indoor temperatures thereby suggesting no change in metabolic rate, however subjects reported significant increase in workload and effort with increased thermal discomfort. Other detailed research in Reference [15] studied the effects of cold temperature on cognitive performance, wherein they observed three distinct performance patterns—negative, positive and mixed, which were determined based on accuracy, response time and efficiency based on a cognitive test battery. They concluded that skin temperature, thermal sensation, diastolic blood pressure and heart rate were independent predictors of decreased accuracy and response time and concluded that cold temperatures impact performance negatively due to mechanisms of distraction and arousal. These past studies indicate performance trends change depending on the task and environmental conditions, which is not always straightforward. More research evidence suggests that human performance is a byproduct of psychological and physiological factors collectively, which we theorize may be better explained by neurophysiological signals.

Taking into account the relationship between human performance and indoor thermal conditions and the advantages of predicting performance by potential improvements on office-workers' health and productivity, we propose to use neurophysiological signals from electroencephalography (EEG) as predictors of performance. Over time, EEG research has been extensively used and shown to be effective in the detection and interpretation of brain mental states during the execution of cognitive and physical tasks. Specifically, the association of cognitive functions with specific brain regions and their temporal characteristics have been determined from imaging studies such as functional magnetic resonance imaging (fMRI), evoked response potential (ERP) and time-frequency analyses of EEG or EEG-MEG (magnetoencephalography) studies. Working memory studies have shown theta band (4–8 Hz) power is correlated with cognitive performance; high-performing individuals or individuals with working memory training exhibit increased theta power in the frontal-parietal brain network [16]. Studies in References [17–21] have also reported the functional involvement of the frontal-parietal network associated with working memory and executive functions and

References [22,23] have reported the involvement of the frontal-temporal declarative and semantic memory network associated with controlled retrieval of task-relevant facts or rules. Researchers have also analyzed the temporal dynamics of these networks; time-frequency analysis in Reference [17] during arithmetic problem-solving tasks shows the engagement of the frontal cortex at around 300 ms from stimulus presentation for memory retrieval strategies reflected as enhanced theta power within the frontal-temporal network. On the other hand, procedural strategies have higher execution demands at later time points, reflected as alpha power event-related desynchronization (ERD) in the frontal-parietal networks.

Analogous to arithmetic problem-solving tasks, brain dynamics have also been reported in tasks involving motor movements where focused attention and somatosensory information processing play a crucial role [24–26]. Tasks that involve motor movements are associated with the activation of contralateral sensorimotor cortex, where findings by References [27–29] report an increase in theta power localized at the fronto-midline during the onset or preset of a motor movement particularly during high performance or by expert performers and increased theta power was additionally observed during higher workloads. Beta (14–30 Hz) oscillations have been known to be associated with voluntary movements, particularly, beta modulations post-movement synchronization over the sensorimotor cortex has been linked to greater confidence in the execution of motor tasks suggesting reinforcement of the current motor state and generation of the steady motor output [30–33]. Beta modulation has additionally been linked to reaction time where a decrease in beta was observed upon committing an error resulting in longer reaction times for upcoming trials due to increased cognitive load [34].

Based on the evidence stated above, establishing performance changes under varying environmental conditions and linkage between behavioral changes/performance with underlying brain activities, we propose to use EEG brain signals to predict performance. With this goal, we present an experimental design wherein subjects perform mental tasks under varying thermal conditions and develop linear regression models to predict performance using EEG power spectral densities (PSD) as features/regressors. Specifically, we theorize the involvement of theta power from the frontal-temporal or frontal-parietal network in arithmetic problem-solving and the involvement of theta and beta/alpha power band from the fronto-midline and motor-cortex for typing tasks to vary at different performance levels. Both office-work tasks in this study require crucial physiological factors such as sustained attention, working memory, self-motivation and motor control specific to typing tasks. To achieve our goal, we first compute the prediction strength of features such as thermal survey scores, heart rate and skin temperature and then compare them to prediction accuracies using EEG power spectral densities from linear regression models. Given the spatial-temporal brain dynamics to complete the task, we implement least absolute shrinkage and selection operator (LASSO) as a feature selection technique to select relevant power densities from brain regions contributing towards explaining performance. Lastly, the robustness of these regressors is compared with other non-neurophysiological signals by reporting least mean square errors (MSE).

2. Materials and Methods

2.1. Office-Work Task Simulation

All participants were required to complete two types of office work task—addition and typing—in two different indoor room temperatures, 22.2 °C (72 F) or 30 °C (86 F). Each task lasted for 15 min (30 min in total). The difficulty level of each task ranged from easy to average, designed with the intention of simulating daily office responsibilities. All participants were provided with a training session to familiarize themselves with the experiment setup, task instructions and software interface.

All participants attempted the addition task first, involving the addition of two three-digit numbers, which were generated randomly online in MATLAB [35]. The task was designed to be self-paced and participants were instructed to avoid errors while attempting as many questions as possible in 15 min; thus, the total number of questions answered by each participant depends on their response time for each question. This was followed by 15 min of a typing task, in which all participants were instructed to type the paragraph (4 sentences long) exactly as presented on the display monitor and was self-paced. The writing paragraph for this task was selected from a journal and no limit was posed on the number of paragraphs typed—that is, every time the participant finished typing the current paragraph, a new paragraph was presented. Similar instructions were provided, that is, to avoid any typing errors and to attempt typing as many paragraphs as possible. The typing software continuously monitored the typed words for errors, in which case the participant had to correct them before proceeding to the next word. Contrary to the typing task, wherein the participant is aware of typing errors and must correct them, in the addition task the participant is unaware of their response accuracy, that is, no feedback was provided.

MATLAB [35] was used to design and program the addition task presentation and the typing task software was developed by the National Research Council of Canada [36].

2.2. Participants

Seven healthy male adult participants, all university students, were recruited for this study whose age ranged between 18 and 25 years (mean age = 23.5 ± 0.8 years). All provided written consent to participate in the study, which was approved by the Institution Review Board at the University of Texas, San Antonio and stated that they were healthy, without any neurological issues and were not under the influence of any drugs at the time of the experiment. All participants reported to have at least five hours of sleep the night before the experiment and dressed in formal casuals (jeans with long sleeve shirts) for the experiment. This clothing level was selected to keep the participants thermally neutral at room temperature 22.2 °C (72 F), which is reported as a neutral temperature to achieve optimal performance. The study was conducted in an experiment room simulating an office environment with comfortable lighting. Each participant was exposed to two thermal conditions—22.2 °C (72 F) and 30 °C (86 F). A ventilation rate of 6 L/s per person was kept constant at both room temperatures and the relative humidity in room was maintained within normal recommended limits. Lastly, before beginning the experiment all participants were instructed to focus and not to move their head or talk during the task.

2.3. Experimental Procedure

First, all participants were guided to a preparation room where a neutral temperature of 22.2 °C was maintained. Here, participants were prepped for the experiment, that is, sensors for measuring skin temperature, heart rate and the EEG cap were attached. After which, they were guided to the experiment-room, the room temperature was randomly maintained at either 22.2 °C or 30 °C, see Figure 1. All participants were seated on a comfortable chair 50 cm away, from the center of the monitor to the participant's eye. Before the start of the first office-work task under each exposure (or session), 10 min of rest time was provided to adapt to the thermal settings and all participants were alone in the experiment-room. Prior to the second exposure, a 45-min break was given to relax, drink water, walk around and use the restroom. In the meantime, the temperature of the experiment-room was increased or decreased depending on the temperature setting used in the first exposure. The order of the indoor room temperature was randomized for each participant, wherein 4 participants were first exposed to 22.2 °C and the remaining three participants to 30 °C. In the second session, all participants repeated the office-work task for the next 30 min. Additionally, before and after each session and each task, participants answered a short thermal survey. The entire experiment lasted for 155 min.

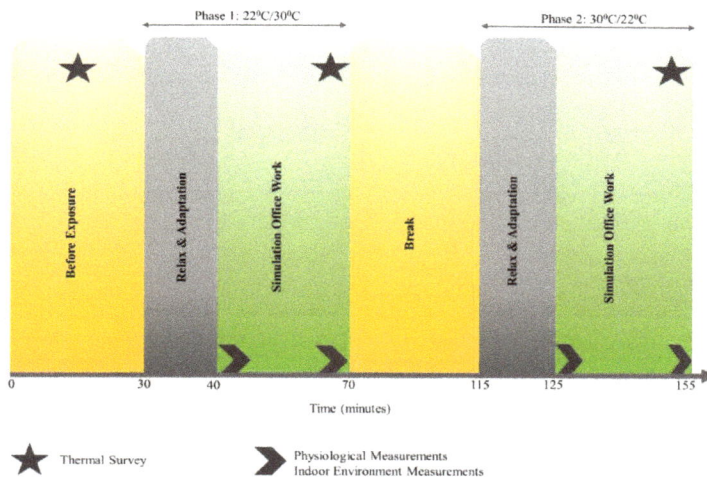

Figure 1. Illustration of experiment timeline.

2.4. Measurements

2.4.1. Performance Metrics

All participants performed the addition and typing task for 15 min under each exposure. Response time and accuracy are the performance metrics commonly used for the addition task. To assess the overall performance in this task, the two metrics were integrated, that is, the time taken to complete 20 questions correctly, that is Equation (1):

$$\begin{aligned} &Addition\ Performance\ Index\ (API)\\ &= Time\ taken\ (seconds)\ to\ answer\ 20\ questions\ correctly \end{aligned} \tag{1}$$

To increase the number of samples, a sliding window of 20 correct questions with a shift of one question is applied, moving along the dimension of number of questions answered. For instance, if the first 20 questions are all answered correctly, then the API for the first sample is calculated as the sum of the response times for answering the first 20 questions. Now, if question number 21 is incorrect but number 22 is correct, then the API of the second sample is calculated as the sum of the response times for answering questions 2 to 22 including exactly 20 correctly answered questions. Thus, for committing an error, a penalty in time is issued in the metric API. We chose 20 correct questions in the metric because most participants take approximately one minute to answer 20 questions, thus making API a stable metric to assess addition performance.

The metric used to evaluate the typing task performance is net characters per minute (CPM) [20], which is calculated as Equation (2):

$$\begin{aligned} &Net\ characters\ per\ minute\ (CPM)\\ &= Total\ number\ of\ key\\ &- (Total\ cursor\ keys\ pressed + 2 \times Number\ of\ backspace\ keys\ pressed) \end{aligned} \tag{2}$$

During the task, the user types the paragraph displayed on the screen. The text is confirmed after each word and in the case of errors a strikethrough is notified on the screen from the point of error occurrence. The user is unable to continue typing until the error has been rectified. The user is unable to use the mouse, however can move around the screen using cursor keys and can delete using BACKSPACE or DELETE keys. The typing performance metric is calculated as the net number of characters typed per minute as shown above. The backspace key is doubled as characters typed are deleted and then retyped. Thus, as the typing errors increase, the number of characters typed per minute (or typing performance) decreases. To calculate CPM samples, a sliding window of one minute was applied with a shift of 30 s.

2.4.2. Physical Measurements

The temperature and relative humidity of the experiment-room were continuously maintained and recorded with data loggers—temperature (range: 20 °C to 70 °C, accuracy: ± 0.7 °C), humidity (range: 0–95%, accuracy: $\pm 5\%$) and CO_2 (range: 0–2000 ppm, accuracy: ± 50 ppm) sensors. All sensors were calibrated before use.

Subjective measurements: A survey/questionnaire was provided to all participants before each task to assess the room thermal conditions (comfort and sensation) and air-quality. The perceived thermal comfort and sensation conditions were assessed using continuous scales describing participants' satisfaction in the thermal environment. In case of thermal comfort, participants reported their comfort level in the room temperature under an exposure. A score of one-point indicates very uncomfortable, 4-point indicates just right and seven-point indicates very comfortable. Likewise, for thermal sensation, one-point indicates cold, 4-point indicates neutral and seven-point indicates hot body sensation. In addition to these questions, participants also answered questions indicating their general indoor thermal preference and if they preferred the current room temperature to be changed.

2.4.3. Physiological Measurements

The physiological measurements included: (1) skin temperature measured from eight sensors located at forehead, right scapula, left upper chest, wrist, both upper arms, left hand, left-calf and right anterior thigh according to ISO 9886 standards. Samples were recorded every second and for analysis purposes a weighted average skin temperature was computed, recommended by ISO 9886 standards [37]; (2) Heart rate was measured by using Polar H7 Smart Chest Transmitter (Polar Electro Oy, Kempele, Finland) and recorded on an iPad via Bluetooth every second.

2.4.4. EEG Measurement and Preprocessing

Brain activities were continuously recorded at a sampling rate of 512 Hz using 64-channel EEG system (Biosemi, Inc. [38]) referenced to the right and left ear mastoids based on a modified international 10–20 system. Before data acquisition, care was taken to ensure that the impedance between EEG electrodes and cortex was less than 5 kΩ. From each participant, 30-min EEG signals during each exposure were recorded and preprocessed prior to obtaining power spectral density (PSD) values for further analysis. EEG preprocessing involved down-sampling the data to 128 Hz, bad channel removal and interpolation using the software EEGLab [39], referencing each EEG electrode using the average signal from left and right ear mastoid connections, bandpass filtered between 1 and 50 Hz to remove electrical noise, DC shift and artefact removal introduced by eye blinks and muscle movements. EEG data from each participant from both exposures were normalized using z-scores. Preprocessing was followed by average PSD value computation for each EEG electrode data epoch. Length of the epochs depended on the type of office work task metric, for the addition task, the length of epochs was based on the time taken to answer 20 questions correctly from its metric API and for the typing task, an epoch length of one minute was extracted based on its metric net CPM. To increase the sample size, a sliding window was applied, wherein for the addition task, a sliding window of 20 questions with one question shift was applied and for the typing task, a sliding window of one minute with a 30-min shift was applied.

3. Results and Discussion

The goal of this paper is to assess the efficiency of using EEG signals in performance prediction induced by varying indoor room temperatures. To do so, this paper is organized into three parts:

first, we present statistical results to validate performance is effected by indoor temperatures; second, we show the prediction results of office-work performance using features reported by past research—thermal sensation, thermal comfort, skin temperature and heart rate in a linear regression model; and third, we present the prediction ability and robustness of using EEG PSDs as predictors in a linear regression model enhanced with LASSO.

3.1. Performance versus Room Temperatures

Tables 1 and 2 summarize the statistical test results of all seven subjects during each office-work task to determine change in performance under different indoor temperatures. For each task, the average performance of the corresponding task is reported under each temperature exposure along with standard deviation in parentheses and respective p-values. Additionally, prior to and after the experiment, all participants answered a thermal survey reporting their comfort levels at 22.2 °C and 30 °C and most felt comfortable at 22.2 °C, which is considered the control exposure in our study design.

Table 1. Kolmogorov-Smirnov (KS) test results on addition task performance under two indoor temperatures. Columns 2 & 3 show the average task performance with standard deviation in parenthesis. Column 4 shows the p-values from the statistical test.

Subject	22.2 °C (72 F)	30 °C (86 F)	KS-Test (p-Value)
S1	86.9 (±8.5)	101.9 (±18.8)	7.2741×10^{-15}
S2	75.9 (±5.5)	70.1 (±8.6)	3.8359×10^{-15}
S3	85.9 (±13.3)	87.0 (±7.5)	0.0051
S4	69.5 (±7.8)	64.5 (±5.3)	3.7328×10^{-11}
S5	99.7 (±9.9)	90.0 (±6.5)	1.2082×10^{-18}
S6	73.5 (±8.5)	78.6 (±7.9)	6.4751×10^{-11}
S7	90.1 (±11.9)	93.5 (±15.3)	0.0021

Table 2. KS test results on typing task performance under two indoor temperatures. Columns 2 & 3 show the average performance with standard deviation in parenthesis. Column 4 shows the p-values from the statistical test.

Subject	22.2 °C (72 F)	30 °C (86 F)	KS-Test (p-Value)
S1	185.25 (±27.3)	207.5 (±25.9)	0.0186
S2	121.5 (±18.8)	122.17 (±34.1)	0.1687
S3	240.6 (±22.8)	226.7 (±30.8)	0.0875
S4	199.7 (±26.2)	214.5 (±16.2)	0.0076
S5	104.5 (±27.9)	120.5 (±19.3)	0.0420
S6	178.3 (±37.3)	191.3 (±26.9)	0.1687
S7	228.0 (±28.1)	244.9 (±37.3)	0.0420

The Kolmogorov-Smirnov (KS) statistical test was used because performance values for both office-work tasks did not follow a normal distribution. In the addition task, the samples used for the KS-test were the time taken to answer 20 questions correctly with a sliding window with an overlap of 19 questions, thus low performance corresponds to a longer time taken to answer 20 questions. The test revealed that all seven participants showed significant differences in performance between the two exposures (p-value < 0.1). As expected, we observed that four out of seven participantsshowed low performance at elevated temperature of 30 °C. In theseparticipants, an increase in response time to answer the arithmetic problems could be attributed to fatigue thereby requiring higher cognitive demand. In the typing task, KS-test samples used were the net characters typed per minute with a sliding window of one minute with an overlap of 30 s, thus fewer characters per minute reflects low performance. Five participantsout of seven showed significant differences in performance between the two exposures, among whom, interestingly, four subjects performed higher in the elevated room temperature of 30 °C. Based on participantfeedback, this is attributed to discomfort at the elevated temperature and thus wanting to finish the task quickly. Although we would expect to observe an increase in typing errors with increased typing speed, this was perhaps not the case because the task was self-paced. Based on these statistical results, we can conclude that indoor room temperature affects office-work performance and increaseor decrease of performance under different room temperatures is task dependent.

3.2. Performance versus Physiological Signals

Table 3 shows the correlations between office-work performance using features reported by past research groups, that is, from thermal survey votes (thermal sensation, thermal comfort) and physiological recordings (skin temperature and heart rate). Data samples used to compute correlations (R^2) between performance and physiological recordings during both tasks are as described in Section 2.4, with the implementation of sliding window for all participants. To compute correlations with thermal sensation and comfort, survey scores were collected at the end of each office-work task (see Figure 1) and corresponding average task performance from all participants were used, without sliding window. Empirical R^2 results show that all predictors exhibit a correlation less than 0.5 ranging between 0.003 and 0.1, indicating that each individual regressor is unable to explain variance in office-work performance and do not exhibit a linear trend. Past research studies have reported correlations of heart-rate variability with mental effort due to its association with blood pressure regulation [40,41]; however, linear correlation analysis in this case did not show significant R^2 correlations. Due to the small and elusive nature of R^2 values, we proceed to investigate the correlation of office-work performance using brain signals obtained from EEG.

Table 3. Correlation R^2 between simulated office-work performance and different physiological predictors.

R^2	Thermal Sensation	Thermal Comfort	Skin Temperature	Heart Rate
Addition Task	0.00369	0.018	0.0127	0.0089
Typing Task	0.0714	0.104	0.0201	0.052

3.3. Performance versus EEG

To investigate the efficiency of predicting performance using EEG power spectral densities in linear regression analysis, we first present results using brain spectral densities as features (i.e., theta, alpha, beta and combined brain bands) from each EEG electrode location separately and then present prediction ability by using a variable selection technique—LASSO—that determines the most relevant features across brain regions.

3.3.1. Band Powers of Raw EEG Data as Regressors

Based on the motivations mentioned above to use EEG brain PSDs as features, we investigate the average spectral powers corresponding to these well-studied oscillations in theta band (4–8 Hz), alpha band (8–14 Hz) and beta band (14–30 Hz).

To study the correlations between change in office-work task performance and EEG spectral bands, a linear regression R^2 analysis was used to determine the relationship between performance and each EEG PSDs from each of the 64 EEG electrode locations from all participants. Figure 2 shows the topoplots of R^2 values obtained for each channel from the above-mentioned three frequency bands. In both office-work tasks, we observed insufficient R^2 ranging approximately between 0.10 and 0.25 when brain power bands are used as regressors individually. Furthermore, we investigated the correlation coefficients (ρ) between pairs of brain power bands corresponding to channels with maximum R^2 in the single regressor linear model. From Table 4, we observe that the correlation coefficients between two regressors is insufficient (<0.9) demonstrating that they do not have a strong correlation, therefore denoting that each PSD contributes independently towards performance prediction. Based on this finding, it is possible to achieve higher correlations R^2 by combining all three brain power bands within each EEG channel to generate a new multiple regressor linear model. In doing so, we observe a maximum correlation of $R^2 = 0.2866$ and $R^2 = 0.3216$ in the addition and typing tasks, each of which is an increase of 21.70% and 26.66% compared to the highest R^2 using a single regressor linear model from each EEG channel. The R^2 topoplots show maximum correlation in the left parietal and occipital brain regions in the addition task and in the right fronto-temporal brain regions in the typing task.

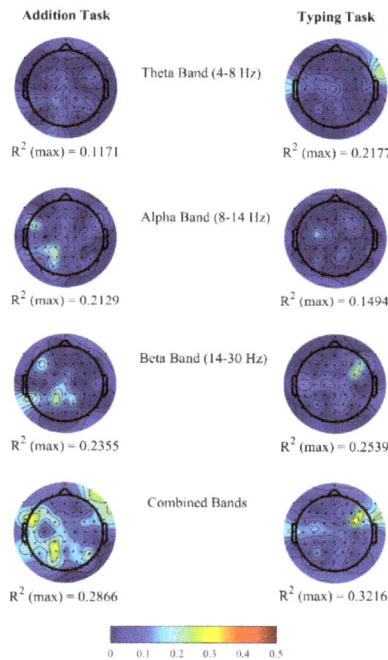

Figure 2. The topoplots represent the correlation R^2 maps between the brain power spectral densities and office task performance.

Table 4. Correlation coefficients (ρ) between brain band pairs corresponding to the EEG electrodes with highest R^2 single regressor linear models.

Single Regressors	Correlation Coefficient (ρ)	Addition Task	Typing Task
Theta Band	Theta–Alpha	0.6978	0.4380
	Theta–Beta	0.6663	0.5549
Alpha Band	Alpha–Theta	0.6788	0.3862
	Alpha–Beta	0.6303	0.6130
Beta Band	Beta–Theta	0.7048	0.3065
	Beta–Alpha	0.6303	0.5176

To ensure the maximum correlation observed is not due to chance/noise permutation, the test was performed by randomizing epochs across EEG channels. p-value = 0 (<0.05) was obtained for all band power regressors, individually and combined, from EEG channel locations corresponding to its maximum R^2 from the linear regression models. Analysis thus far supports the notion that there are multiple brain regions contributing towards an explanation of performance and helping to achieve higher prediction. With this motivation to achieve higher prediction power, we proceed to using LASSO (least absolute shrinkage and selection operator) to select relevant brain bands from specific EEG channels collectively in the next section.

3.3.2. LASSO with Brain Band Power in All Brain Regions as Regressors

To determine the subset of EEG spatial locations that collectively contribute to predicting performance, the LASSO regression analysis method was implemented. This method fits a sparse linear regression model that performs both feature selection to avoid multicollinearities and overfitting regularization to improve prediction accuracies and interpretability of statistical models. Specifically, let $y \in \Re^{N \times 1}$ represent a vector of performance values from N epochs (N = 2546 for addition task and 330 for typing task) and $X \in \Re^{N \times M}$ be a matrix of power values of a frequency band, whose nm^{th} element denotes the power of epoch n at channel m (note when all brain bands are combined $M = 64 \times 3 = 192$). LASSO fits a linear model between y and X given by Equations (3) and (4):

$$y = X\beta + \beta_o + \varepsilon \tag{3}$$

where $\beta \in \Re^{M \times 1}$ and β_o are model coefficients and ε is the $N \times 1$ noise vector with zero mean and constant variance. LASSO aims to find estimates of the coefficients $\hat{\beta}$ by optimizing

$$\min_{\beta_o, \beta} \frac{1}{N} ||y - X\beta - \beta_o|| + \lambda |\beta| \tag{4}$$

where $|.|$ and $|| \; . \; ||$ denote the l_1-norm and l_2-norm respectively and λ is the regularization parameter [42]. The l_1-norm constraint (2) forces the coefficients to be sparse, that is, only small subsets of coefficients are nonzero. There are two advantages of using LASSO that generate sparse constraints. First, as the dimension of the matrix X in (1) is $M = 64$ or 192, representing the number of EEG channels when PSDs are used individually and combined and $N = 2546$ and 330 performance samples from the addition and typing tasks, LASSO avoids overfitting. Secondly, the sparse coefficients make the model more interpretable, as the model focuses only on the powers from channels with nonzero coefficients. To further reduce the overfitting during model fitting, 20% of data samples were set aside for testing—called holdout data—and the remaining was used for model training. LASSO iteratively generates models with different regularization parameters on the training data, after which holdout data is used to determine a model that gives the lowest mean square error between the observed and predicted performance. To estimate the prediction ability of chosen model, R^2 is computed between the observed and estimated office-work performance.

Table 5 summarizes the correlation results between observed and estimated office-work performance from the features selected by LASSO regression model for each office-work task, including the number of EEG electrodes selected by LASSO and *p*-value corresponding to the statistical significance of the model chosen. Overall, the resulting LASSO models are relatively sparse, exhibiting R^2 in the range of 0.64–0.89 and, as expected, is two times greater than the maximum R^2 obtained from the linear regressor model, that is, without combining spatial information, from both office-work tasks. In the addition task, we observe that all PSD regressors, individual and combined, provide correlations >0.5, with the highest R^2 observed using alpha power as the regressor individually and combined with other power bands, using LASSO, at 88.6% from 51 electrode locations and 83.4% from 174 electrode locations. On comparing both R^2 values, using alpha power alone, the LASSO model outperforms the latter by ~5% with contributions from fewer EEG channels (51 EEG channels), maybe by removing channels involved in multicollinearities. In the typing task, R^2 from the LASSO model using a theta power band provides the best performance prediction at 74.6% with contributions from 48 EEG channels. Thus, we conclude that, for the addition task, we use alpha brain power from 51 EEG channels, and for the typing task, we use theta brain power from 48 EEG channels as features for performance predictions.

Table 5. R^2 obtained between observed and estimated performance for office tasks using LASSO regression and the number of non-zero coefficients in the fitted LASSO model.

R^2	Theta Band (4–8 Hz)	Alpha Band (8–14 Hz)	Beta Band (14–30 Hz)	Combined Bands
Addition Task	0.681	0.886	0.67	0.834
(# non-zero coefficients)	(#64)	(#51)	(#62)	(#174)
p-values on fitting	0	0	0	0
Typing Task	0.746	0.712	0.696	0.645
(# non-zero coefficients)	(#48)	(#38)	(#43)	(#45)
p-values on fitting	3.5292×10^{-91}	7.2004×10^{-86}	1.054×10^{-84}	1.6241×10^{-25}

Furthermore, based on the LASSO models obtained for the two office-work tasks we statistically determined the reliability of performance prediction induced by two indoor temperatures with alpha power (i.e., addition task) and theta power (i.e., typing task). In other words, if the residual errors between observed and predicted performance follow a random normal distribution, this indicates that the LASSO model has considered all features in linear regression analyses towards predicting performance. To do so, we use the *t*-test on the error values from both exposures for each task. The null hypothesis for the t-test being performance errors induced by the two room temperatures are the same. *p*-values of 0.7756 and 0.5605 were obtained for the addition and the typing task. At a significance level $\alpha = 0.1$, the tests failed to reject the null hypothesis of equal performance error means, implying that changes in performance are sufficiently explained by the features selected by the LASSO technique.

3.3.3. Prediction of Performance

Finally, we investigated the power of using brain power bands as regressors for predicting performance and compared them to other physiological signals, that is, skin temperature and heart rate. For physiological signals, polynomial models of model orders 1–9, were fitted by using the same data as used for LASSO. Table 6 shows the mean squared errors (MSE) of all biomarkers presented in this paper. It is not surprising to find that the LASSO predictors using PSDs obtained much smaller MSEs than those from skin temperature and heart rate (even with a higher order polynomial model). Taken together, these results confirm that neurophysiological signals recorded using EEG are better predictors of human performance induced by different indoor room temperatures.

Table 6. MSE obtained from LASSO model using brain PSDs & from polynomial curve fitting models using physiological signals.

Brain Band	Mean Square Errors	
	Addition Tasks	Typing Tasks
Theta (4–8 Hz)	79.97	600.42
Alpha (8–14 Hz)	27.55	682.48
Beta (14–30 Hz)	79.15	717.30
Combined Bands	40.15	1127.30
Skin temperature	2612.5 (6th order)	37002 (5th order)
Heart Rate	284.5460 (7th order)	33361 (4sh order)

3.3.4. Brain Activity Pattern

To gain insight into the mental functions that are induced by varying thermal conditions during these office-work tasks, we investigate the differences in brain activity patterns arising from change in performance from all power bands. To do so we categorized participants' performance (i.e., API and net CPM) into two groups, low- and high-performance by defining cutoff values based on the scatter plots obtained between observed and predicted performance using the LASSO model as seen in Figure 3A,B. For the addition task's performance cutoff values, samples less than 100 s were labeled as high performance and values greater than 120 s were labeled as low performance. Likewise, for the typing task, samples less than 100 net CPM were labeled as low performance and those greater than 150 CPM were labelled as high performance. Based on these cutoff values, we plot the average brain activities on the scalp projected from EEG channel locations with non-zero coefficients obtained from the LASSO model from each brain power band, shown in Figures 4 and 5.

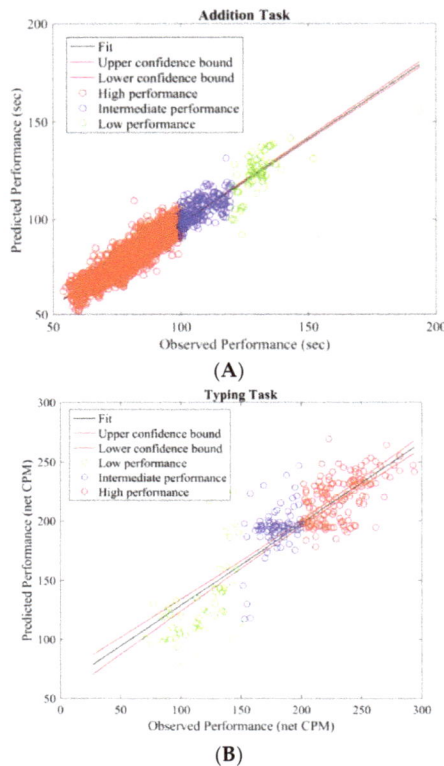

Figure 3. (**A**) Addition task—scatter plot of observed versus predicted performance (seconds) by using LASSO linear regression model with alpha power as a single regressor; (**B**) Typing task—scatter plot of observed versus predicted performance (net CPM) by using LASSO linear regression model with theta power as a single regressor.

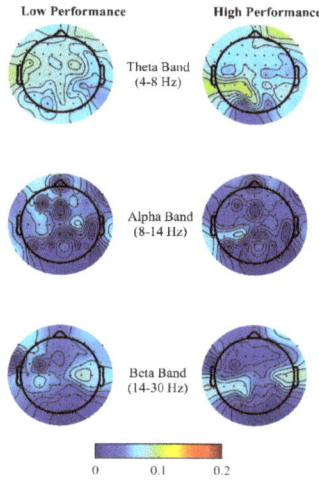

Figure 4. Scalp activities across EEG channels with non-zero LASSO coefficients from brain spectral power for low- and high-addition task performance.

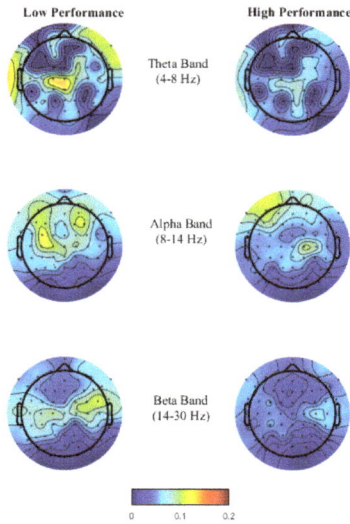

Figure 5. Scalp activities across EEG channels from non-zero LASSO coefficients from brain spectral power for low- and high-typing task performance.

Brain activity patterns from both office-work tasks over low and high performance may be interpreted as the average spatial temporal brain power density patterns over a time window corresponding to the performance index of the specific task, that is, in the addition task, over the time taken to answer 20 questions correctly and for the typing task, net characters typed per minute. For the addition task, see Figure 4, congruent with the findings from References [16–21] we observe high localized theta power over the right prefrontal and left parietal cortex during high performance than compared to low performance. This is indicative of the frontal-parietal network being associated with working memory and executive functions of arithmetic problem-solving. High theta activity at the right frontal cortex reflects arithmetic fact retrieval, while differential brain temporal dynamics across the frontal-parietal cortex reflect higher cognitive demands for multistep procedural strategies. Since frontal theta activity is common in retrieving basic arithmetic facts, LASSO was unable to pick this feature as most discriminant. Also, bilateral beta activity, particularly during high performance, is reported to be associated with the conceptual processing of numbers, that is, the identification of operands in the addition task [43,44]. Although in this study design we are unable to conclusively compare arithmetic problems between retrieval or multistep procedures based on the subjective difficulty levels, we can infer that discomfort due to thermal conditions created demands of higher cognitive load to either maintain the current performance level or to achieve high performance. Overall, in agreement with the findings from Reference [16], desynchronization of alpha and beta power is lower at the frontal and parietal region, with right frontal theta enhancement reflecting distinct cognitive functions in multicomponent problem solving.

High performance in the typing task requires both motor control with sustained attention, a similar psychological requirement to brain oscillations in sports activities. As theorized earlier in the typing task, see Figure 5, theta power in the fronto-midline is found to be the most discriminant feature for performance prediction wherein high theta is observed during low performance at the frontal-midline and the second most discriminant feature is frontal alpha event-related desynchronization (ERD) during high performance. Studies by References [27–29] have associated high frontal theta and high parietal alpha power differentiating skilled sports players to novices, reflecting developed task solving strategies, focused attention and an economic parietal sensory information processing. The results found in this paper for the typing task observed an opposite trend in the theta band at different performance levels, as the typing task involved reinforcement learning where subjects were required to rectify typing errors in order to proceed, forcing subjects to refocus and retype. Thus, enhanced frontal-midline theta power during low performance in our data possibly reflects error feedback information processing and subsequently increasing response time to retype correctly reflected as high beta power at the somatosensory cortex during low performance. Alpha activity amplitudes have been shown to be inversely related to the amount of neuronal population activated during cognitive-motor tasks. Studies by References [45–48] have related alpha and beta ERD for skilled performers to be associated with fine cognitive-motor performance. This is consistent with our findings, where during high performance alpha and beta ERD were observed over premotor and sensorimotor areas reflecting confidence in typing correctly, which requires precise planning and regulation of bilateral finger movements.

Overall, brain activity patterns presented in Figures 4 and 5 show that during high performance there is lower activity than during low performance, which is almost in line with the "neural efficiency" hypothesis. These brain activity patterns enable the creation of a unique EEG profile for varying degrees of performance level, which are task-dependent. In this study, we use EEG sensor space features to predict performance, which are limited by volume conductance, while it is possible that source space estimates could provide better predictions. Additionally, it is possible that functional connectivity estimates at source level could provide higher prediction ability as reported in Reference [49]. However, these are the current two limitations of our study as we analyze data in EEG sensor space only and treat changes in brain states as a continuous task rather than 'event related.' The main motivation to conduct analyses in this fashion is to use this analysis technique in real-time brain computer interfaces in a realistic office-work setting to predict environmental conditions based on performance predictions from EEG power spectral densities.

4. Conclusions

To the best of our knowledge, we are the first to present preliminary results of using EEG power spectral densities to predict performance changes due to change in indoor-room temperature. Our analysis statistically validates that office-work performance is impacted by varying indoor temperature. We present a comprehensive regression analysis for predicting performance in two different office-work tasks using features reported by past studies and neurophysiological EEG signals. We found that EEG brain band power is the best predictor of performance, which was enhanced using the LASSO regression technique. This method found alpha brain power to be the best feature corresponding to the right frontal and left parietal cortex for the arithmetic problem-solving task ($R^2 = 88.6\%$) and theta brain power as the best feature corresponding to the fronto-middle cortex for the typing task ($R^2 = 74.6\%$). Lastly, the robustness of using EEG power spectral densities as features was reported by mean-square errors. With LASSO, we were able to achieve performance prediction abilities five times greater than using a single linear regression model and 17 times higher prediction ability than compared to using thermal survey votes, skin temperature and heart-rate. While the results of this study are promising, there are a few limitations. Currently, we are unable to confirm the behavioral trend, that is, whether there is an increase or decrease in performance under different indoor temperatures due to insufficient population size, which is why we report prediction strength using linear regression analysis and were still able to achieve promising results. Also, it is possible that upon collection of data from more subjects, the relationship between EEG power spectral densities and office-work performance under different thermal conditions may not be linear, thereby a non-linear regression technique, or other machine learning techniques may be needed for classifications. In the future, this research needs to focus on more comprehensive investigations of performance under longer exposures and using varying workloads and further methodological studies are needed to investigate prediction models to classify cross-task performance. Ultimately, this domain of research aims to provide motivation for future research to help achieve optimal productivity from office-workers by providing feedback regarding their environmental conditions.

Acknowledgments: This work is partially funded by the Army Research Laboratory under Cooperative Agreement Number W911NF-10-2-0022.

Author Contributions: T.N. analyzed the data collected and wrote the paper. T.Z and Z.M. collected the data. X.X. designed the experiments. L.Z. processed the data. D.J.P. conceived the idea. B.D. conceived the idea and designed the experiments. Y.H. conceived the idea and wrote the paper.

Conflicts of Interest: The authors declare no conflict of interest.

References

1. Clements-Croome, D. *Creating the Productive Workplace*; Taylor & Francis: Abingdon-on-Thames, UK, 2006.
2. Lan, L.; Lian, Z.; Pan, L.; Ye, Q. Neurobehavioral approach for evaluation of office workers' productivity: The effects of room temperature. *Build. Environ.* **2009**, *44*, 1578–1588. [CrossRef]
3. Wargocki, P. Productivity and health effects of high indoor air quality. In *Encyclopedia of Environmental Health*; Elsevier: Burlington, MA, USA, 2011.
4. Wargocki, P.; Wyon, D.P. The effects of moderately raised classroom temperatures and classroom ventilation rate on the performance of schoolwork by children (RP-1257). *HVACR Res.* **2007**, *13*, 193–220. [CrossRef]
5. Wyon, D.P.; Wargocki, P. *Room Temperature Effects on Office Work*; Taylor & Francis: London, UK, 2006.
6. Fisk, W.J.; Rosenfeld, A.H. Estimates of improved productivity and health from better indoor environments. *Indoor Air* **1997**, *7*, 158–172. [CrossRef]
7. Link, R.P.J. Associated fluctuations in daily temperature, productivity and absenteeism. *ASHRAE Trans.* **1970**, *76*, 326–337.
8. Niemelä, R.; Hannula, M.; Rautio, S.; Reijula, K.; Railio, J. The effect of air temperature on labour productivity in call centres—A case study. *Energy Build.* **2002**, *34*, 759–764. [CrossRef]
9. Federspiel, C.C.; Fisk, W.J.; Price, P.N.; Liu, G.; Faulkner, D.; DiBartolomeo, D.L.; Sullivan, D.P.; Lahiff, M. Worker performance and ventilation in a call center: Analyses of work performance data for registered nurses. *Indoor Air* **2004**, *14*, 41–50. [CrossRef] [PubMed]
10. Wargocki, P.; Wyon, D. Effects of HVAC on Student Performance. *ASHRAE J.* **2006**, *48*, 12.
11. Wyon, D.P.; Andersen, I.; Lundqvist, G.R. The effects of moderate heat stress on mental performance. *Scand. J. Work Environ. Health* **1979**, *5*, 352–361. [CrossRef] [PubMed]

12. Lan, L.; Wargocki, P.; Wyon, D.P.; Lian, Z. Effects of thermal discomfort in an office on perceived air quality, SBS symptoms, physiological responses and human performance. *Indoor Air* **2011**, *21*, 376–390. [CrossRef] [PubMed]

13. Tanabe, S.-I.; Nishihara, N.; Haneda, M. Indoor temperature, productivity and fatigue in office tasks. *HVACR Res.* **2007**, *13*, 623–633. [CrossRef]

14. Wargocki, P.; Delewski, M.; Haneda, M. Physiological effects of thermal environment on office work. *Healthy Build.* **2009**, *2*, 1270.

15. Mäkinen, T.M.; Palinkas, L.A.; Reeves, D.L.; Pääkkönen, T.; Rintamäki, H.; Leppäluoto, J.; Hassi, J. Effect of repeated exposures to cold on cognitive performance in humans. *Physiol. Behav.* **2006**, *87*, 166–176. [CrossRef] [PubMed]

16. Langer, N.; von Bastian, C.C.; Wirz, H.; Oberauer, K.; Jäncke, L. The effects of working memory training on functional brain network efficiency. *Cortex* **2013**, *49*, 2424–2438. [CrossRef] [PubMed]

17. Tschentscher, N.; Hauk, O. Frontal and parietal cortices show different spatiotemporal dynamics across problem-solving stages. *J. Cogn. Neurosci.* **2016**, *28*, 1098–1110. [CrossRef] [PubMed]

18. Cole, M.W.; Reynolds, J.R.; Power, J.D.; Repovs, G.; Anticevic, A.; Braver, T.S. Multi-task connectivity reveals flexible hubs for adaptive task control. *Nat. Neurosci.* **2013**, *16*, 1348. [CrossRef] [PubMed]

19. Duncan, J. The multiple-demand (MD) system of the primate brain: Mental programs for intelligent behaviour. *Trends Cogn. Sci.* **2010**, *14*, 172–179. [CrossRef] [PubMed]

20. Dosenbach, N.U.; Fair, D.A.; Cohen, A.L.; Schlaggar, B.L.; Petersen, S.E. A dual-networks architecture of top-down control. *Trends Cogn. Sci.* **2008**, *12*, 99–105. [CrossRef] [PubMed]

21. Sammer, G.; Blecker, C.; Gebhardt, H.; Bischoff, M.; Stark, R.; Morgen, K.; Vaitl, D. Relationship between regional hemodynamic activity and simultaneously recorded EEG-theta associated with mental arithmetic-induced workload. *Hum. Brain Mapp.* **2007**, *28*, 793–803. [CrossRef] [PubMed]

22. Barredo, J.; Öztekin, I.; Badre, D. Ventral fronto-temporal pathway supporting cognitive control of episodic memory retrieval. *Cereb. Cortex* **2013**, *25*, 1004–1019. [CrossRef] [PubMed]

23. Badre, D.; Wagner, A.D. Left ventrolateral prefrontal cortex and the cognitive control of memory. *Neuropsychologia* **2007**, *45*, 2883–2901. [CrossRef] [PubMed]

24. Cheron, G.; Petit, G.; Cheron, J.; Leroy, A.; Cebolla, A.; Cevallos, C.; Petieau, M.; Hoellinger, T.; Zarka, D.; Clarinval, A.M.; et al. Brain oscillations in sport: Toward EEG biomarkers of performance. *Front. Psychol.* **2016**, *7*, 246. [CrossRef] [PubMed]

25. Thompson, T.; Steffert, T.; Ros, T.; Leach, J.; Gruzelier, J. EEG applications for sport and performance. *Methods* **2008**, *45*, 279–288. [CrossRef] [PubMed]

26. Del Percio, C.; Babiloni, C.; Bertollo, M.; Marzano, N.; Iacoboni, M.; Infarinato, F.; Lizio, R.; Stocchi, M.; Robazza, C.; Cibelli, G.; et al. Visuo-attentional and sensorimotor alpha rhythms are related to visuo-motor performance in athletes. *Hum. Brain Mapp.* **2009**, *30*, 3527–3540. [CrossRef] [PubMed]

27. Baumeister, J.; Reinecke, K.; Liesen, H.; Weiss, M. Cortical activity of skilled performance in a complex sports related motor task. *Eur. J. Appl. Physiol.* **2008**, *104*, 625. [CrossRef] [PubMed]

28. Doppelmayr, M.; Finkenzeller, T.; Sauseng, P. Frontal midline theta in the pre-shot phase of rifle shooting: Differences between experts and novices. *Neuropsychologia* **2008**, *46*, 1463–1467. [CrossRef] [PubMed]

29. Ofori, E.; Coombes, S.A.; Vaillancourt, D.E. 3D Cortical electrophysiology of ballistic upper limb movement in humans. *Neuroimage* **2015**, *115*, 30–41. [CrossRef] [PubMed]

30. Fischer, P.; Tan, H.; Pogosyan, A.; Brown, P. High post-movement parietal low-beta power during rhythmic tapping facilitates performance in a stop task. *Eur. J. Neurosci.* **2016**, *44*, 2202–2213. [CrossRef] [PubMed]

31. Tan, H.; Jenkinson, N.; Brown, P. Dynamic neural correlates of motor error monitoring and adaptation during trial-to-trial learning. *J. Neurosci.* **2014**, *34*, 5678–5688. [CrossRef] [PubMed]

32. Tan, H.; Zavala, B.; Pogosyan, A.; Ashkan, K.; Zrinzo, L.; Foltynie, T.; Limousin, P.; Brown, P. Human subthalamic nucleus in movement error detection and its evaluation during visuomotor adaptation. *J. Neurosci.* **2014**, *34*, 16744–16754. [CrossRef] [PubMed]

33. Tan, H.; Wade, C.; Brown, P. Post-movement beta activity in sensorimotor cortex indexes confidence in the estimations from internal models. *J. Neurosci.* **2016**, *36*, 1516–1528. [CrossRef] [PubMed]

34. Jha, A.; Nachev, P.; Barnes, G.; Husain, M.; Brown, P.; Litvak, V. The frontal control of stopping. *Cereb. Cortex* **2015**, *25*, 4392–4406. [CrossRef] [PubMed]

35. Guide, M.U.S. *The Mathworks*; MathWorks Inc.: Natick, MA, USA, 1998; Volume 5, p. 333.

36. Newsham, G.; Veitch, J.; Scovil, C. *Typing Task: Software to Measure the Speed and Accuracy with Which Presented Text Is Typed*; National Research Council Canada: Ottawa, ON, Canada, 1995.

37. International Standard Organization. *Ergonomics-Evaluation of Thermal Strain by Physiological Measurements*, 2nd ed.; International Standard Organization: Geneva, Switzerland, 2004; pp. 1–21.

38. BioSemi, B.V. *BioSemi ActiveTwo.[EEG System]*; BioSemi: Amsterdam, The Netherlands, 2011.

39. Delorme, A.; Makeig, S. EEGLAB: An open source toolbox for analysis of single-trial EEG dynamics including independent component analysis. *J. Neurosci. Methods* **2004**, *134*, 9–21. [CrossRef] [PubMed]

40. Paas, F.G.; van Merriënboer, J.J.; Adam, J.J. Measurement of cognitive load in instructional research. *Percept. Motor Skills* **1994**, *79*, 419–430. [CrossRef] [PubMed]

41. Thayer, J.F.; Hansen, A.L.; Saus-Rose, E.; Johnsen, B.H. Heart rate variability, prefrontal neural function and cognitive performance: The neurovisceral integration perspective on self-regulation, adaptation and health. *Ann. Behav. Med.* **2009**, *37*, 141–153. [CrossRef] [PubMed]

42. Tibshirani, R. Regression shrinkage and selection via the lasso. *J. R. Stat. Soc. Ser. B* **1996**, *58*, 267–288.

43. Berryhill, M.E.; Olson, I.R. Is the posterior parietal lobe involved in working memory retrieval?: Evidence from patients with bilateral parietal lobe damage. *Neuropsychologia* **2008**, *46*, 1775–1786. [CrossRef] [PubMed]

44. Cappelletti, M.; Lee, H.L.; Freeman, E.D.; Price, C.J. The role of right and left parietal lobes in the conceptual processing of numbers. *J. Cogn. Neurosci.* **2010**, *22*, 331–346. [CrossRef] [PubMed]

45. Babiloni, C.; Del Percio, C.; Iacoboni, M.; Infarinato, F.; Lizio, R.; Marzano, N.; Crespi, G.; Dassù, F.; Pirritano, M.; Gallamini, M.; et al. Golf putt outcomes are predicted by sensorimotor cerebral EEG rhythms. *J. Physiol.* **2008**, *586*, 131–139. [CrossRef] [PubMed]

46. Klimesch, W. EEG alpha and theta oscillations reflect cognitive and memory performance: A review and analysis. *Brain Res. Rev.* **1999**, *29*, 169–195. [CrossRef]

47. Klimesch, W.; Doppelmayr, M.; Pachinger, T.; Ripper, B. Brain oscillations and human memory: EEG correlates in the upper alpha and theta band. *Neurosci. Lett.* **1997**, *238*, 9–12. [CrossRef]

48. Pfurtscheller, G.; da Silva, F.L. Event-related EEG/MEG synchronization and desynchronization: Basic principles. *Clin. Neurophysiol.* **1999**, *110*, 1842–1857. [CrossRef]

49. Dimitrakopoulos, G.N.; Kakkos, I.; Dai, Z.; Lim, J.; Bezerianos, A.; Sun, Y. Task-Independent Mental Workload Classification Based Upon Common Multiband EEG Cortical Connectivity. *IEEE Trans. Neural Syst. Rehabil. Eng.* **2017**, *25*, 1940–1949. [CrossRef] [PubMed]

MDPI

St. Alban-Anlage 66

4052 Basel

Switzerland

Tel. +41 61 683 77 34

Fax +41 61 302 89 18

www.mdpi.com

Brain Sciences Editorial Office

E-mail: brainsci@mdpi.com

www.mdpi.com/journal/brainsci

www.ingramcontent.com/pod-product-compliance
Lightning Source LLC
Chambersburg PA
CBHW051911210326
41597CB00033B/6115